Recent Trends in
INVERTEBRATES

Vol. 4

H.S.Bhamrah
Kavita Juneja

ANMOL PUBLICATIONS

ANMOL PUBLICATIONS
4378/4B, Ansari Road,
Darva Ganj, New Delhi - 110002.

Recent Trends in Invertebrates

ISBN 81 7041 473 3 (Set)
ISBN 81 7041 474 1 (Vol.I)
ISBN 81 7041 475 X (Vol.II)
ISBN 81 7041 476 8 (Vol. III)
ISBN 81 7041 477 6 (Vol. IV)
ISBN 81 7041 478 4 (Vol. V)
ISBN 81 7041 479 2 (Vol. VI)
ISBN 81 7041 480 6 (Vol. VII)
ISBN 81 7041 481 4 (V0l. VIII)

PRINTED IN INDIA

Published by J.L. Kumar for Anmol Publications and Laser Typeset by D.V. Computers, Delhi, Printed at Efficient Offset Printers, New Delhi.

PREFACE

The present title has been designed to meet the requirements of the undergraduate students. In the Recent Trends in Invertebrate all phyla have been discussed. The representative types of each phylum have been described in detail. This is followed by the classification of the phylum with a description of its common animals. Recent findings and work have been incorporated in the text. Special care has been taken to make the language as simple and lucid as possible. Simple and accurate diagrams accompany the text. In the interest of the students efforts have been made to cover the syllabi of nearly all Indian Universities.

The book features both a text and a laboratory guide. It deals with the structures and physiological phenomenon of each system. General topics are also dealt in details. It is hoped that this book will not only meet the requirement of undergraduate students but also be useful as a guideline to the teachers in their teaching. It is also hoped that the general public will find that the book is not riddled with too much of scientific and technical jargon.

It is difficult to acknowledge adequately the assistance of all who have contributed to the preparation of this book. First of all the authors express their thanks to their friends and colleagues whose continuous inspiration have initiated them to bring out the present volume.

The authors are grateful to Dr. J. Bahadur, Professor and Head, School of Studies in Zoology, Jiwaji University Gwalior, Dr. A.K. Pandey, Professor and Head, School of Studied in Zoology, Vikaram University, Ujjain for their valuable suggestions.

Constructure criticism from students and teachers will be highly appreciated by the authors.

H.S. Bhamrah

Kavita Juneja

CONTENTS

1

HELMINTHES

Helminthes constitute a large assemblage of 'worms' of comparatively simple and varied organization. Helminthes is derived from Greek word *helmins* means 'worms'. The term worms is not correct because it is applied to elongated invertebrates without appendages and with bilateral symmetry. In 18th century all the worm-like animals were included in an old phylum-*Vermes* (*Linnaeus*, 1735). But this was a very heterogenous group which included a highly diverse assemblage of forms. In the modern classification this group is divided into several independent phyla such as Platyhelminthes, Aschelminthes, Annelida etc.

The term helminthes is restricted to a few phyla of invertebrate animals, all of which are superficially worm-like but they differ markedly in their morphology, life-history and bionomics. The helminthes are restricted to two phyla of Annual Kingdom. These are :

1. Platyhelminthes or flat-worms
2. Asc helminthes or round-worms

Phylum - Platyhelminthes

Term *Platyhelminthes* was proposed by *Gegenbaur* in 1859, means flatworms which refers to their characteristic contour of flatenned body. The term is derived from two Greek words - *platys* = flat ; *helminthes* = worms. The animals show low organisation as they are without anus, skeletal, respiratory and circulatory system. The body is filled with mesenchymal cells which are mesodermal in origin.

General Characters

1. Bilateral symmetrical body, with dorso-ventral flattening.
2. Body generally shows anterior cephalization that is deferentiation of anterior region into head.
3. Body generally worm like either long flat ribbon shaped or leaf-like. The outline of body bizaree shape in Trematodes and segmented tape-like body in Cestodes.
4. Most of the flat worms are of small to moderate dimensions so size varies from microscopic to the extreme elongation as much as 10 to 15 mm.
5. General colour creamy white but on account of the presence of food the body may acquire colour. Some of the free-living flat worms are brilliantly coloured, often brown, black or gray.
6. Body with various types of attachment in the form of adhesive secretions, suckers and hooks.
7. Body consists of three embryonic layer viz. ectoderm, mesoderm and endoderm.
8. In parasitic forms, the body is covered with a thick cuticle but free-living Turbellarians are clothed with cellular or syncytial epidermis.
9. The exo and endoskeleton are completely absent therefore the body is generally soft with hooks, spines, spicules and thorns.
10. Body space is filled with a mesenchyma that is parenchymatous tissues, therefore, flat worms are accoelomate that is without coelom.
11. The digestive system is generally absent in Tapeworms and Acoela. But in other consists of mouth, pharynx and variously modified intestine.
12. Respiratory and circulatory systems are absent but some of the trematodes have system of tubes the so called lymphatic system of uncertain function.
13. The nervous system is of primitive type with missing of nervous tissue as brain in anterior-most region with several longitudinal ganglionated cord. Numerous transverse connections occur between the longitudinal cords. So the nervous system is generally ladder type.
14. Parasitic forms are without sensory organs but free-living forms are provided with sensory organs in the form of chemo-and tango receptors ciliated pits, statocysts and ocelli or eyes.

15. Excretory system well developed with proto-nephridial tubules with flame bulbs or cells. But accoela are without protonephridia.
16. Hermaphroditerarely with separate sex. The eggs are generally devoid of yolk but with yolk cells. Varieties of copulatory organs are present. Cross fertilization and self-fertilization both are common.
17. Development direct i.e. without larval stage or indirect with larval stages. Life cycles complicated with succession of larval stages involving one to three intermediate host.
18. Asexual reproduction common among fresh water forms.
19. Parthenogenesis and polyembryony common among trematodes and cestodes.
20. Endogenous and exogenous budding common in tapeworms.

Phylum platyhelminthes is divided into three classes. The classification followed here is from *Hyman* (1951).

I Class - Turbellaria
II Class - Trematoda
III Class - Cestoda

I CLASS - TURBELLARIA

Turbellaria is derived from Latin word *turbella* means a *stirring.* The animal produce turbulance in water by body cilia. They possess following characters:

1. The members of this class are broad, flat and leaf-like in shape.
2. They are lead a free living life in fresh or salt water, in moist soil, some members may be commensal or parasite.
3. Body unsegmented, dorso-ventrally flatenned and covered with cilia.
4. The anterior end is differentiated into head.
5. Epidermis cellular or syncytial contains rod-like structures called rhabdites. The rhabdites are curved structure capable of discharging a chemical secretion.
6. Epiderrmis and mesenchyma are provided with unicellular glands of two types (a) cyanophilous which secrete adhesive

secretion and (b) eosinophilous.

7. Turbellarians are abundently provided with adhesive organs which are of two types (a) glandulo - epidermal adhesive organs and (b) glandulo - muscular adhesive organs.
8. Alimentary canal incomplete i.e. with single opening, the mouth. Mouth is mid-ventral, sometimes near the centre of body. Intestine is blind or absent.
9. Excretory system with protonephridia.
10. Sense organs consist of tangoreceptors and chaemoreceptors.
11. Mostly bisexual or hermaphrodite or unisexnal with very few exceptions.
12. Reproduction by asexual, sexual and by regeneration.
13. Cleavage spiral. Development direct without larval stage.

Classification of Turbellaria

The members of this class are found almost all places. It includes more than 1500 species class Turbellaria is divided into 5 orders.

1. Acoela
2. Rhabdocoela
3. Allocoela
4. Tricladida
5. Polycladida

Order 1 - Acoela (Gr. a = without Koilos = cavity)

1. They are minute, delicate turbellarians with a ventral mouth.
2. Mouth and pharynx are present. Pharynx simple when present but usually absent and so the intestine is also absent.
3. Parenchyma syncytial not differentiated into mesoderm and endoderm.
4. Protonephridia absent.
5. Food is taken in solid form and digested by endodermal cells.
6. Gonoducts, gonads and yolk glands are absent.
7. Exclusively marine forms living under stores algae etc. at the sea bottom. Some live in the intestine of holothurians.
   Examples : Convoluta, Hoplodiscus

Order 2 - Rhabdocoela (Gr. rhabdos = rod ; Koils = cavity)

1. Small Tubellarian usually less than 3mm in length.
2. Mostly free-living, marine or fresh water.

Some of them inhabit in the respiratory cavities of various fresh-water crustaceans, *viz. Temnocephala.*

3. Body is elongated, cylindrical or compressed with simple pharynx and simple unbranched sac-like intestine.
4. Yolk-glands may or may not be present.
5. Male with two to many testes, whereas the female with 1 to 2 ovaries.
6. Nervous system consists of two longitudinal nerve cords. Eyes are present in most of them.

Order Rhabdocoela is divided into four sub-orders:

Sub-order (i) Notandropora

1. Exclusively fresh water forms.
2. Pharynx simple.
3. Single median protonephridium.
4. Single compact testis mass, penis unarmed.
5. Yolk-gland is absent.
6. Asexual fission with chains of zooids.
   Examples : *Catenula, Rhycoscolex, Stenostomum*

Sub-order (ii) Opisthandropora

1. Fresh water or marine forms.
2. Pharynx is simple.
3. Excretory system consists of paired protonephridia.
4. Compact testes, penis armed with a stylet.
5. Yolk glands absent.
6. Asexual reproduction with chains of zooids.
   Examples: *Microstomum, Macrostomum*

Sub-order (iii) Lecithophora

1. Fresh water, marine or terrestrial, commensel or parasitic.
2. Pharynx bulbose.
3. Protonephridia paired.
4. Separate ovaries and yolk glands.
5. Exclusively sexual reproduction.
   Examples: *Graffilla, Anaplodium, Merostoma, Gyratrix*

Sub-order (iv) Temnocephalida

1. Freshwater ectocommensal forms.
2. Anterior end having 2 - 12 tentacles.
3. Posterior end with 1 or 2 adhesive discs.
4. Pharynx dolii form.
5. Gonopore simple.
   Examples: *Temnocephala, Monodiscus*

Order 3 - Allocoela (Gr. Alloios = different ; Koilos = cavity)

1. Mostly marine, few freshwater and brakish water forms. Some are ectoparisite or ectocommensal in habit.
2. The pharynx is bulb-like-or plicate.
3. Intestine is straight or branched.
4. Excretory system consists of paired protonephridia with two or three branches and nephridiopores.
5. Nervous system with 3 or 4 pairs of longitudinal nerve cords provided with transverse commissures.
6. Reproductive system consists of numerous testes and a pair of ovaries. Penis papilla mostly present.

Order Allocoela is divided into four Sub-orders.

Sub-order (i) Archiophora

1. Marine animals.
2. Pharynx plicate.
3. Female reproductive system primitive.
   Examples : *Proporoplana*

Suborder (ii) Lecithoepitheliata

1. Marine, freshwater or terrestrial forms.
2. Pharynx simple or bulbose.
3. Penis with a cuticular stylet.
4. Female duct simple or none. Typical yolk glands absent, nutritive cells may surround the ova.
   Examples : *Hofstenia, Prorhynchus.*

Suborder (iii) Cumulata

1. Freshwater or marine forms.
2. Pharynx plicate or bulbose.
3. Intestine mostly without diverticulae.

4. Penis unarmed.
5. Separate ovaries and yolk glands or *germoritellaria.*
Examples : *Hydrolimax, Plagiostomum.*

Suborder (iv) Seriata

1. Mostly marine and a few are freshwater forms.
2. Pharynx plicate.
3. Intestine with lateral diverticula.
4. Statocyst mainly present.
5. Ovaries and yolk glands are separate.
Examples : *Monocelis, Otoplana*

Order 4 Tricladila (Gr. treis = three ; Klados = branch)

1. Marine, freshwater or terrestrial forms.
2. Large turbellarians, measuring 2-60 cm in length.
3. Pharynx plicate usually backwardly directed.
4. Intestine three-branched sacs.
5. Eyespots usually present.
6. Two ovaries, 2 or numerous testes.
7. Female reproductive system provided with one or two copulatory bursa.
8. Penis with penis papilla, common genital aperture.

Order Tricladida is divided into three suborders.

Suborder (i) Maricola

1. Exclusively marine forms.
2. A pair of eyes and auricular groores.
3. Typical penis papilla sometimes armed with stylet.
4. Rounded copulatory bursa is generally present.
5. Only sexual reproduction.
Examples : *Bdelloura, Ectoplana*

Suborder (ii) Paludicola

1. Mostly freshwater rarely brackish water forms.
2. Two to many eyes or eyeless.
3. Bursa with bursal canal, anterior to penis.
4. Reproduction mostly asexual.
Examples : *Dugesia, Kenkia, Dendrocoelum*

Suborder (iii) Terricola

1. Terrestrial tropical & subtropical forms.
2. Two or many eyes.
3. Bursa, if present, lies behind the penis.
4. Reproduction may be asexual.
   Examples : *Geoplana, Dolicoplana*

Order 5 - Polycladida (Gr. poly = many ; klados = branch)

1. Size modrate, from 2-20 mm. Dorsoventrally flat leaf-like body; they are exclusively marine.
2. Intestine with many branches.
3. Eyes and nerve cords are present.
4. There are numerous ovaries and testes.
5. Genital apertures separate.

It includes two suborders :

Suborder (i) Acotylea

1. Pharynx usually vertical curtain like.
2. No suckers are present behind the female pore.
3. Tentacles nuchal type.
4. Eyes concentrated at anterior margin, not in pairs.
   Examples : *Shylochus, Notoplana*

Suborder (ii) Cotylea

1. Sucker present behind the female pore.
2. Pharynx tubular.
3. A pair of marginal tentacles bearing eyes or with a cluster of eyes in place of tentacles.
   Examples : *Thysanozoon, Yungia*

CLASS II TREMATODA

1. They are ecto or endoparasites and commonly called 'flukes'.
2. Body is usually unsegmented, dorso-ventrally flatenned and leaf-like.
3. Size varies from 0.5 mm to 1 or several cm. long.
4. Body is covered over by thick cuticle and no cilia.

5. Suckers and sometimes hooks are present. Oral sucker present at anterior end for feeding ventral sucker for attachment.
6. Epidermis absent. Rhabdites are absent.
7. Digestive system consists of mouth, muscular pharynx and intestine usually with two main branches. Anus is absent.
8. In some cases there is a system of parenchymal vessels showing the presence of primitive circulatory system.
9. Excretory organs consists of protonephridia.
10. There are three pairs of longitudinal nerve cords. Sense organs are poorly developed.
11. Bisexual or hermaphrodite except blood fluke (*Schistosoma*).
12. Single ovary and many testes. Cross fertilization takes place. Development direct or indirect.
13. Life-cycle simple or complicated with more than one hosts.

There are about 6000 species in class Trematoda. This class is divided into three orders:

Order 1 - Monogenea
Order 2 - Aspidobothrea
Order 3 - Digenea

Order 1 - Monogenea

1. They are ectoparasite mostly in cold-blooded animals (fishes).
2. The life-cycle is simple and passed in one host only, monogenetic.
3. Oral sucker absent, when present is poorly developed.
4. A pair of adhesive structures are present at anterior end.
5. A large disc-shaped structure, the opisthaptor, is found on the posterior end. It contains hooks and is lined with chitin.
6. Paired excrecory pores are present in the anterior part of the body.
7. Male and female genital pores are separate.
8. One or two vaginae are present. The uterus is small and contains few eggs.
9. Development is indirect through a larval stage called onchomiracidium.

Order Monogenea is divided in two suborders:

Suborder (i) Monopisthocotylea

1. Opisthaptors are present without suckers but with hooks.
2. Genito-intestinal tract is absent.
3. Posterior adhesive organ is single or divided by septa.
4. Eye-spots usually present.
   Examples : *Gyrodactylus, Monocotyle, Heterocotyle, Amphibdella*

Suborder (ii) Polyopisthocotyle

1. Opisthaptor contains many suckers and also hooks.
2. Oral suckers two, opening into oral cavity.
3. Genito-intestinal tract present.
4. Eye-spots are absent.
   Examples : *Polystoma, Diplozoon, Microcotyle.*

Order 2 - Aspidobothrea

1. Endoparasite in the digestive tract of fishes and reptiles.
2. Oral sucker is absent.
3. The ventral sucker is divided by septa into many small parts. The hooks are absent in it.
4. Single excretory pore.
5. Only one testis is present.
6. Development direct, without alternation of hosts.

Order Aspidobothrea is divided into two families:

Family (a) Aspidogastridae

1. Sucker enormous circular, oval or elongated, divided by septa into one, three or four longitudinal rows of depressions termed alveloli.
2. Protonepliridia have separate bladders and separate or a common nephridiopore situated near the posterior end.
3. Testis single.
4. Laurer's canal may be present.
5. Endoparasite in the mantle, pericardial and renal cavities of clams and snail and in digestive system of fish and turtle.
   Examples : *Aspidogaster, Cotylaspis, Cotylogaster*

Family (b) Stichocotylidae

1. It consists of only one genus which lives in bile passage or spiral valve of fishes.
2. The life cycle is complex one, involving at least two hosts.
   Example : *Stichocotyle.*

Order 3 - Digenea

1. They are endoparasite in invertebrate and vertebrates.
2. Two to four hosts in the life-cycle.
3. Two suckers, without hooks, mouth within anterior sucker.
4. There is only one excretory pore present in the posterior region of the body.
5. There is a common gonopore for male and female reproductive systems. Vagina is absent.
6. Uterus is long branched with many eggs.
7. Their larval forms reproduce asexually before metamorphosis.

Order Digenea is divided into many families ; some important ones are as :

Family (a) Bucephalidae

1. They are commonly called gastrostomes. The mouth is located in the middle of the body.
2. Anterior end bears a simple or lobate oral sucker or has a generally lobed hood-like extension.
3. Intestine sacciform.
4. Acetabulum absent.
5. Genital pore posterior.
6. The cercaria larvae known as oxhead, have a pair of long tails.
7. The primary host is fish and secondary host is clams.
   Examples : *Bucephalus, Bucephalopsis.*

Family (b) Paramphistomidae

1. Body thick and fleshy with a powerful posterior sucker.
2. Cirrus pouch is present.
3. Oral sucker is absent.
4. Gonopore, situated in the anterior body half and provided with in some genera with a genital sucker.
5. Nephridiopore dorsal infront of acetabulum.
6. Uterus long having ascending limb. Vitelline glands follicular.
7. These are the endoparasites of gut and bile ducts.
   Examples : *Paramphistomum, Gastrothylax*

Family (c) Opisthorchidae

1. These are distomes of small to large size.

2. They live in the bile passage of fish-eating reptiles, birds and mamnials.
3. Body flat and semitransparent.
4. Excretory bladder 'V' shaped.
5. Testes posterior Genital pore infront of acetabulum.
   Examples : *Opisthorchis, Amphimerus, Metorchis*

Family (d) Fasciolidae

1. Commonly known as liver flukes.
2. Body leaf-like and acetabulum near the oral sucker.
3. Gonopore just infront of the acetabulum.
4. Testes, ovary, vitelline glands all are branched.
   Examples : *Fasciola hepatica, Fasciplopsis buski*

Family (e) Echinostomatidae

1. Body slender, elongated with spiny or scaly cuticle.
2. Acetabulum large behind the oral sucker.
3. Collar of spines is present in cercarial and metacercarial stages.
4. It is commonly found in the intestine of birds.
   Examples: *Echinostoma, Echinoparyphium, Parochis, Himasthla*

Family (f) Schistosomatidae

1. These are blood flukes that inhabit the hepatic portal system and pelvic veins of birds and mammals.
2. Sexes are separate.
3. Female lodges in a gynaecophoric canal of male.
4. Testes numerous and ovary tubular.
   Example : *Schistosoma*

Besides above families the others are :

Family - Zoognidae.	Examples: *Zoogonus, Nassa.*
Family - Allocreadidae.	Examples: *Bunodera, Plagioporus.*
Family - Azygiidae.	Examples: *Azygia, Otodistomum.*
Family - Hemiuridae.	Examples: *Halipegus, Derogenes.*
Family - Didymozoonidae.	Examples: *Didymocystis*
Family - Gorgoderidae.	Examples: *Gorgoderina, Gorgodera*
Family - Plagiorchidae.	Examples: *Plagiorchis, Styphlodora.*
Family - Dicrocoelidae.	Examples: *Dicrocoelium, Eurytrema.*
Family - Heterophyidae.	Examples: *Stictodera, Heterophyes.*

Family - Microphallidae.	Examples: *Spelotrema, Maritrema.*
Family - Troglotrematidae.	Examples: *Paragonimus, Troglotrema.*
Family - Cyclocoelidae.	Examples: *Cyclocoelus, Typhlocoelum.*
Family - Clinostomatidae.	Example : *Clinostomum.*
Family - Notocotylidae.	Examples: *Notocotylus, Catatropis.*
Family - Strigeidae.	Examples: *Ophiosoma, Cyanthocotyle*
Family - Brachylaimidae.	Examples: *Brachylaima.*
Family - Sanguinicolidae.	Example : *Sanguinicola.*
Family - Spirorchidae.	Examples: *Spirorchis, Hapalorhynchus*
Family - Heronimidae.	Example: *Heronimus.*

CLASS III CESTODA

1. They are found as endoparasites in the intestine of vertebrates and are called tapeworms.
2. Body is flat, elongated and ribbon-like, measures from 1 mm to 10 metres in length.
3. Body is without epidermis, rhabdites, cilia but a cuticle is present.
4. Body is usually divided into a scolex or head, neck and few to many proglottids.
5. Scolex is provided with hooks or suckers or both.
6. Scolex is followed by neck which forms new proglottids by transverse budding called strobilization.
7. Mouth and digestive tract are absent.
8. Excretory system consists of protonephridia with flame - cells.
9. Nervous system consists of ganglionated nerve ring in scolex and two pairs of lateral nerve cords in proglottids.
10. Each proglottid contains one or two sets of male and female reproductive organs. Thus they are hermaphrodite.
11. Several testes in the form of follicles and single-lobed ovary.
12. Uterus presents a great variety of structures.
13. Fertilization internal and self.
14. Life-cycle is complicated with hooked embryo, and is passed in more than one host.

Class Cestoda has about 3400 species. The class is divided into two sub-classes:

Sub-class (A) Cestodaria or Monozoa
Sub-class (B) Eucestoda or Merozoa

Sub-class (A) Cestodaria

1. Endoparasites in the intestine and coelomic cavities of fishes and reptiles.
2. Body is formed by one segment and without a scolex.
3. No alimentary canal.
4. Single set of reproductive organs.
5. Larva with ten hooks, known as Pycophore.

Sub-class Cestodaria is divided into two orders.

Order 1 - Amphilinida

1. They are endoparasites in the body cavity of ganoid fishes.
2. Frontal gland and protrusible pharynx are present in the anterior part of the body.
3. Suckers are absent.
4. A proboscis is present.
5. Male and female pores present in the posterior part of body.
6. Uterus is coiled and opens near the anterior end.
   Example : *Amphilina*

Order 2 - Gyrocotylidea

1. They are endoparasites of chimaeroid fishes.
2. Anterior end has cup-like sucker.
3. Eversible proboscis at anterior end.
4. Male and female pores are situated at the anterior half of the body.
5. Uterus not coiled.
   Example : *Gyrocotylus*.

Sub-class (B) Eucestoda

1. Elongated, flat body is divided into many proglottids.
2. Body is divisible into scolex, neck and strobila.
3. Scolex with suckers, spines and hooks for attachment to host.
4. Each proglottid with male and female reproductive organs.
5. Larva with six hooks.

Sub-class Eucestoda is divided into following orders.

Order 1 - Tetraphyllidea

1. Endoparasite, exclusively in the intestine of elasmobranch fishes.
2. Scolex provided with four bothria (sessile suckers) often with hooks.
3. Testes lie infront of ovaries.
4. Cirrus armed with hairs, spines or hooks.
5. Common genital atrium marginal.

   Examples : *Phyllobothrium, Acanthobothrium.*

Order 2 - Lecanicephaloidea

1. Intestinal parasites of elasmobranch fishes.
2. Scolex divided by a transverse groove in two parts upper with disc and lower with four suckers.
3. Vitellaria in two lateral bands.

   Examples : *Lecanicephalum, Polypocephalus*

Order 3 - Proteocephalidea

1. These are endoparasites of freshwater fishes, amphibians and reptiles.
2. Scolex mobile with four lateral suckers, an apical sucker in some.
3. Ovary bilobed, Uterus with many branches and pores.
4. Vitellaria scattered.
5. Common genital atrium marginal.

   Examples : *Proterocephalus, Ophistaenia*

Order 4 - Diphyllidea

1. Parasitic in the intestine of elasmobranch fishes.
2. Proglottids only 20 or less.
3. Scolex with two bothria and a spiny head stalk.
4. Vitellaria scattered.

   Example : *Echinobothrium*

Order 5 - Trypanorhyncha

1. Parasitic in the spiral valve of digestive tract of elasmobranch fishes.
2. Strobila is segmented and is of modrate size.
3. Scolex with two or four sessile bothria and four protrusible spiny proboscis.
4. Testes extend beyond ovary posteriorly.

   Examples : *Haplobothrium, Tetrarhynchus.*

Order 6 - Pseudophyllidea

1. Endoparasites in the intestine of fishes, birds and mammals.
2. The larva lives in crustaceans.
3. Generally six shallow groove-like suckers and hookless bothrio are found on the scolex.
4. Proglottids may be absent or present.
5. Neck is not developed.
6. Many yolk glands are present.

   Examples: *Amphicotyle, Dibothriocephalus*

Order 7 - Nippotaeniidea

1. Endoparasite in the intestine of Japanese freshwater fishes.
2. Scolex with single terminal sucker.
3. Proglottids few with, each with one set of genitalia.
4. Vitellaria compect.

   Examples : *Nippotaenia, Amurotaenia*

Order 8 - Taenioidea

1. Endoparasites in the intestine of birds and mammals.
2. Elongated body, with proglottids.
3. Scolex bears four suckers often with an apical rostellum armed with hooks.
4. Excretory system with four longitudinal vessels and flame cells.
5. Complete sets of reproductive organs in each proglottids. Single compact yolk gland.

   Examples : *Taenia, Echinococcus.*

Order 9 - Aporidea

1. It is endoparasite of swan(birds).
2. Scolex with four suckers, rostellum armed.
3. No segmentation, ootype, yolk glands and sex-ducts or pores.

   Examples : *Nematoparataenia, Gastrotaenia*

POLYCHOERUS

Phylum	-	**Platyhelminthes**
Class	-	**Turbellaria**
Order	-	**Acoela**
Genus	-	***Polychoerus***

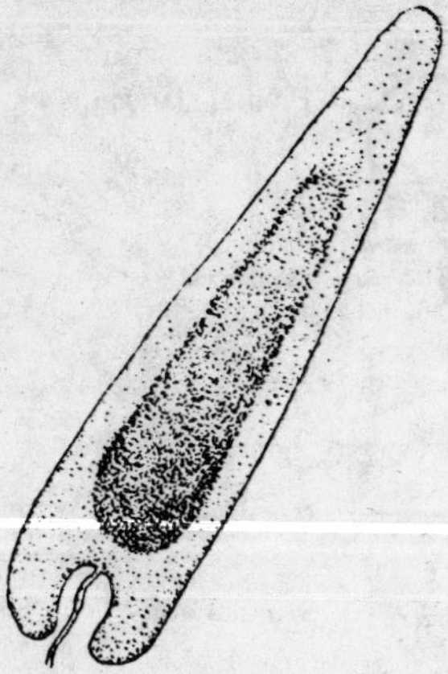

Fig. 1.1. *Polychoerus.*

1. It is found in littoral zone of ocean shores in colder winter.
2. It is bright orange or orange red in colour.
3. Body flat, oval, changeable shape. The anterior end is round while posterior end is not ched bearing tail filament.
4. Entire body is covered with cilia.
5. Statocyst is present in median like.
6. Digestive tract is absent.
7. Definite gonads are wanting. Yolk glands are absent. Definite oviduct absent.
8. The seminal bursa opens ventrally by female gonopore.
9. Penis is present.

CONVOLUTA

Phylum	-	**Platyhelminthes**
Class	-	**Turbellaria**
Order	-	**Acoela**
Genus	-	***Convoluta***

1. Convoluta is found in sea water.
2. Living under stones, among algae and most commonly on the bottom mud.
3. Small worm with side of the body are held curved ventrally without any projections.

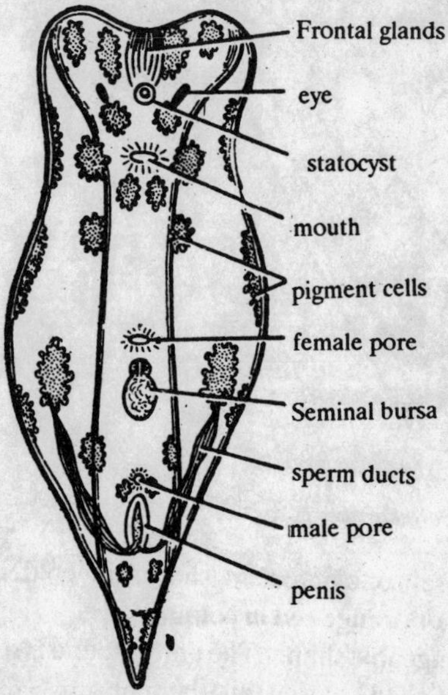

Fig. 1.2 *Convoluta*

4. But the body has pigment cells.
5. The anterior of cephalic end has two eyes, a statocyst and cluster of frontal glands.
6. The mouth is located near the anterior half of the ventral surface.
7. A digestive cavity is lacking.
8. The excretory system is completely absent.
9. Hermaphrodite but protandry is common.
10. The eggs are enclosed in a thin gelatinous capsule and are struck to some objects.
11. The eggs are developed directly into young worms that escape from the jelly.

CENTULA

Phylum	-	**Platyhelminthes**
Class	-	**Turbellaria**
Order	-	**Rhabdocoela**
Suborder	-	**Notandropora**
Genus	-	***Centula***

Fig. 1.3. *Centula.*

1. It is a freshwater Turbellaria.
2. The pharynx is simple.
3. Single median protonephridium.
4. Nervous system consists of four pairs of longitudinal nerve cords and a statocyst.
5. Asexual reproduction by fission with chains of zooids.
6. No sexual reproduction.

MICROSTOMUM

Phylum	-	**Platyhelminthes**
Class	-	**Turbellaria**
Order	-	**Rhabdocoela**
Genus	-	***Microstomum***

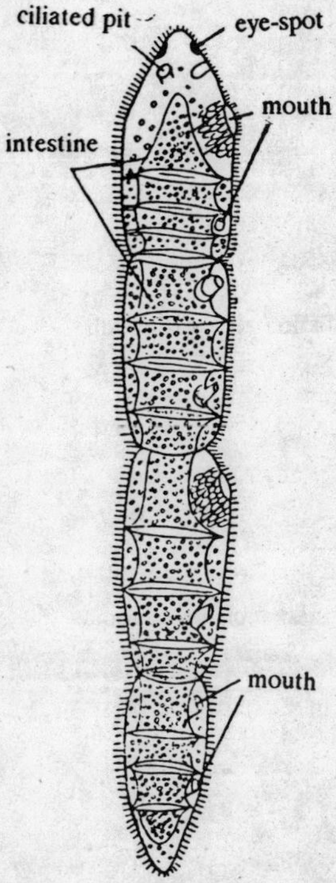

Fig. 1.4 *Microstomum.*

1. It is a microscopic, freshwater form.
2. It swims by means of cilia present on the body surface.
3. Its epidermis contains nematocysts derived from hydroid food.
4. It is found that Microstomum attack Hydra only when it need of nematocysts.
5. The digestive cavity is simple, unbranched and straight.
6. It reproduce asexually by repeated transverse fission forming a linear chain of 4, 8 or 16 zooids. Each zooid separates to become an individual.

TEMNOCEPHALA

Phylum	-	**Platyhelminthes**
Class	-	**Turbellaria**
Order	-	**Rhabdocoela**
Genus	-	***Temnocephala***

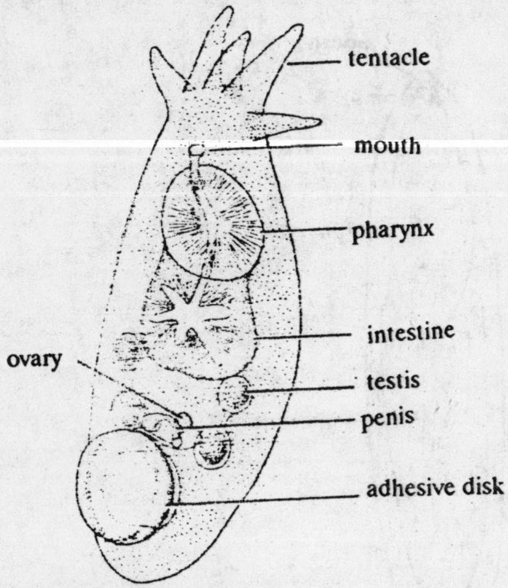

Fig. 1.5 *Temnocephala.*

1. It is found as an ecto commensal on cray fish, prawns and other crustaceans.
2. It contains 12 finger-like projection or *tentacles* at anterior end and an adhesive disc at the posterior end.
3. The border of syncytial epidermis is apparently cuticle devoid of cilia.
4. Mouth anterio-ventral. Pharynx thick walled and simple lobulated intestine present.
5. It feeds upon diatoms, rotifers and other small animals.
6. A pair of nephridiopores is present on the dorsal surface of the anterior part of the body.
7. Cerebral ganglion bears two dorsal eyes.
8. There are four testes and single ovary. Their gonoducts leading into a common genital atrium.

PLAGIOSTOMUM

Phylum	**-**	**Platyhelminthes**
Class	**-**	**Turbellaria**
Order	**-**	**Allocoela**
Genus	**-**	***Plasgiostomum***

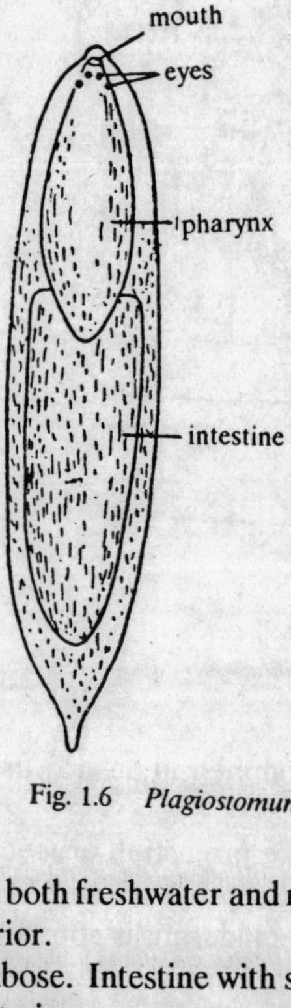

Fig. 1.6 *Plagiostomum.*

1. It is present in both freshwater and marine.
2. Mouth is anterior.
3. Pharynx is bulbose. Intestine with short diverticula.
4. Gonopore posterior.
5. Sperm is characteristic in having wing-like lateral expansions.

BIPALIUM

Phylum	**-**	**Platyhelminthes**
Class	**-**	**Turbellaria**
Order	**-**	**Tricladida**
Genus	**-**	***Bipalium***

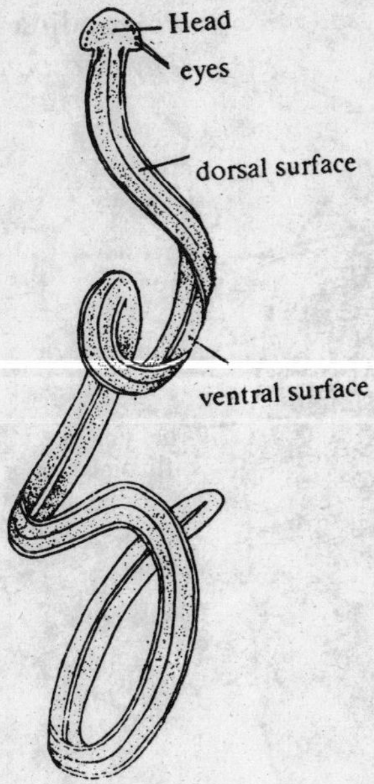

Fig. 1.7. *Bipalium.*

1. Bipalium is a long terrestrial turbellarian living the humid soil on the floor of tropical forests.
2. It is cosmopolitan in distribution.
3. The animal measures 20-50 cm in length.
4. It consists of an expanded lunate head and cylindrical long body.
5. Numerous eyes are present on the margin of the head and sides of body.
6. There are stripes on the dorsal surface and a creeping sole on the ventral surface.
7. The stripes are purple, black, yellow, olive and grey in colour.
8. Reproduction generally asexual. It never becomes sexual in temperate climate and it propogates by fragmentation.
9. Bipalium adventium breeds sexually.

THYSANOZOON

Phylum	-	**Platyhelminthes**
Class	-	**Turbellaria**

Order	-	**Polycladida**
Genus	-	***Thysanozoon***

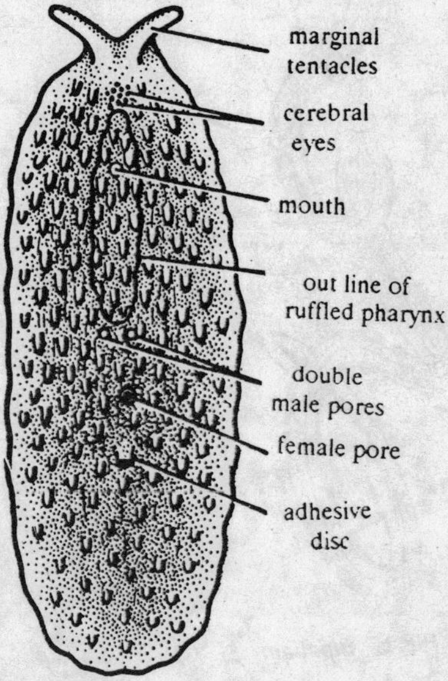

Fig. 1.8 *Thysanozoon.*

1. It is found in a less cold water of sea.
2. It has an oval body with dorsal surface beset with papillae, each receiving a branch from intestine.
3. Anterior end bears a pair of marginal tentacles and numerous cerebral eyes.
4. Pharynx is tubular.
5. Glandulo-muscular adhesive organ is present on the dorsal surface behind the female gonopores.
6. A single female genital pore lies behind the male genital pores.
7. Seminal bursa is absent.

GUNDA

Phylum	-	**Platyhelminthes**
Class	-	**Turbellaria**
Order	-	**Tricladida**
Genus	-	*Gunda* (*Procerodes*)

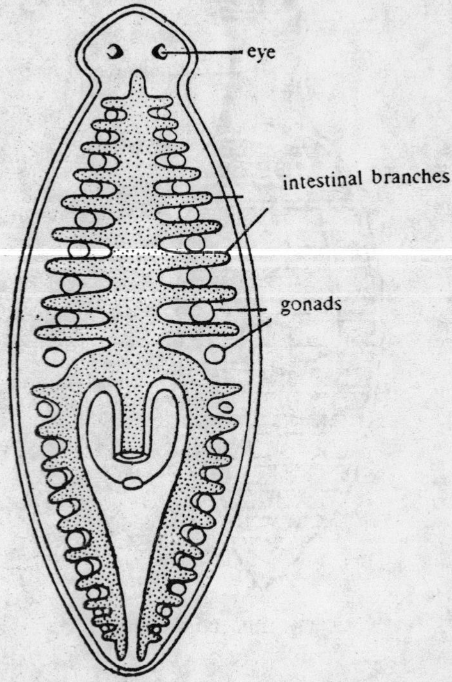

fi.g 1.9 *Gunda segmentina.*

1. It is a marine animal.
2. The auricle and a pair of eye or ocelli are present in the anterior end.
3. Intestine has regularly arranged intestinal diverticulae. The gonads alternate with the diverticulae giving the appearance of segmentation.
4. Incapable of sexual reproduction.
5. *Lang* used the feature of intestine as a classical example in support the gonocoel theory of coelom formation by Haeckel.

APIDOGASTER

Phylum	-	**Platyhelminthes**
Class	-	**Trematoda**
Order	-	**Monogenea**
Genus	-	*Apidogaster*

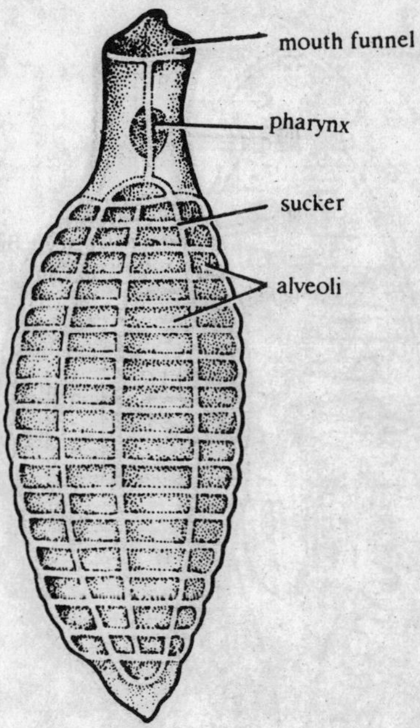

Fig. 1.10 *Aspidogaster.*

1. It is an endoparasite in the pericardial and renal cavities of freshwater mussel and in the gut of fishes and turtles.
2. Body is elongated and dorsoventrally flatenned with anterior narrow end.
3. The narrow anterior end has a subterminal mouth which is devoid of oral sucker.
4. A large sucker appears which occupies great space in the posterior region of body. The sucker is sub-divided into four longitudinal rows of sucking cups or alveoli.
5. Gut is simple and straight.
6. Excretory system consists of protonephridia with excretory bladders.
7. Bisexual life-cycle simple without change of host.

POLYSTOMUM

Phylum - Platyhelminthes
Class - Monogenea

Order - Monogenea

Genus - *Polystomum*

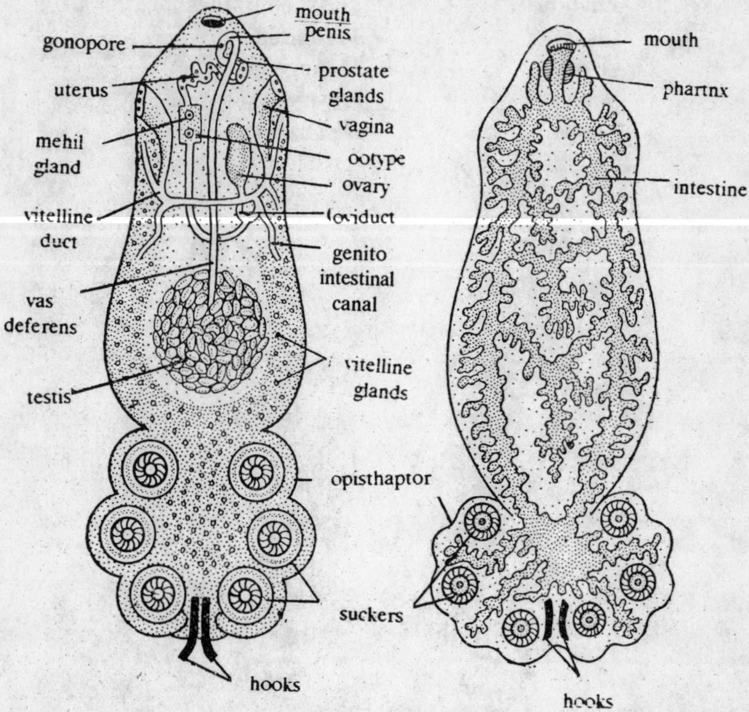

Fig 1.11 *Polystomum.*

Characters:

1. It is an endoparasite in the urinary bladdar of frogs, and turtles.
2. Body is leaf like and dorso-ventally flattened.
3. The posterior region has a disc-shaped opisthaptor with six cup-like suckers and 2 to 4 large chitinous hooks.
4. Mouth is surrounded by a oral sucker.
5. Intestine is bifurcated.
6. Hermaphrodite with single ovary and numerous tests.
7. Breeding season starts in spring. Eggs are produced in the urinary bladder of frog and discharged into the water along with the urine of the host.
8. Hatching takes place after 4 to 5 weeks and larva attaches to the gills of tadpole and later develop into adult in the alimentary canal of metamorphosed tadpole.

DIPLOZOON

Phylum	-	**Platyhelminthes**
Class	-	**Trematoda**
Order	-	**Monogenea**
Suborder	-	**Polyopisthocotyla**
Genus	-	***Diplozoon***

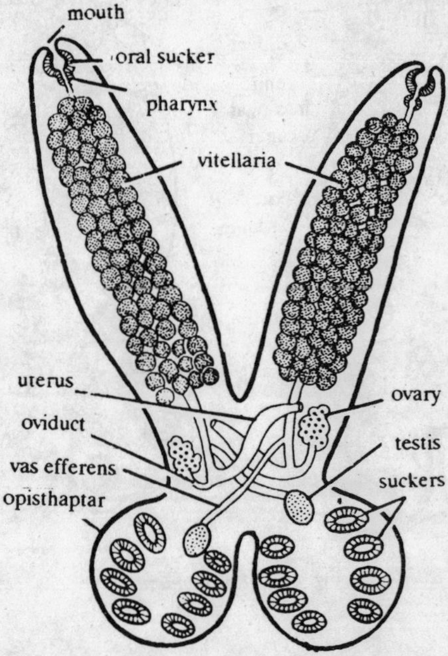

Fig. 1.12 *Diplozoom.* Two individuals.

Characters :

1. Diplozoon is an ectoparasite, found attached to the gills of freshwater fishes, feeding on their blood.
2. Two individuals are always found attached permanently together in the form of 'X'.
3. Opisthaptor bears eight suckers arranged in two rows of four each.
4. Digestive system consists of mouth surrounded by oral sucker, pharynx and intestine. Intestine is not bifurcated.
5. Reproductive system consists of a testis and ovary along with their usual ducts.

6. The sperm duct, uterus and vagina of one individual cross with those or other and thus ensures cross fertilization. Uterus surrounded by mehlis' glands and contains one egg.
7. Egg differenciates into diporpa larva, two of which unite and metamorphosed into adult individual. Larva has two eyes digestive system & a pair of suckers.

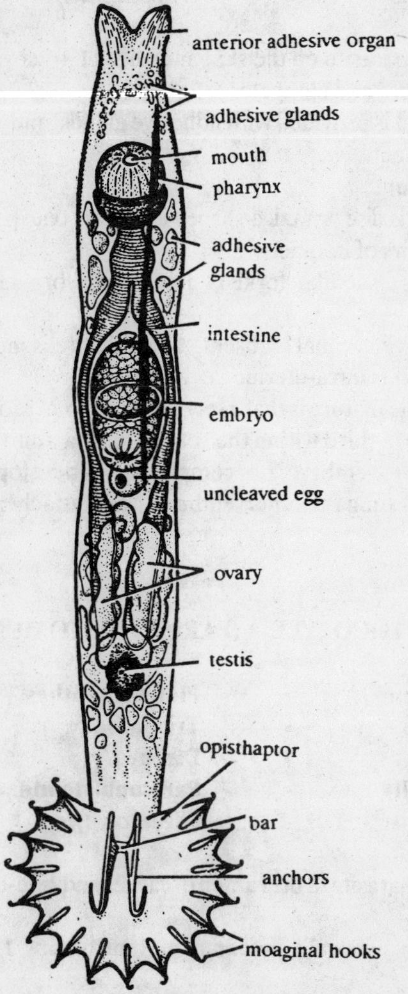

Fig. 1.13 *Gyodactylus.*

GYRODACTYLUS

Phylum	-	**Platyhelminthes**
Class	-	**Trematoda**
Order	-	**Monogenea**
Suborder	-	**Monopisthocotyla**
Genus	-	***Gyrodactylus***

1. It is an ectoparasite on the skin and gills of freshwater fishes.
2. Body is minute and elongate.
3. Anterior end is provided with adhesive glands and organs. Oral sucker is absent.
4. Eyes are absent.
5. Opisthaptor is disc or wedge-shaped and bears one pair of anchors and eight pairs of hooklets.
6. Intestine is sac-like forked into two branches without diverticula.
7. Genito-intestinal canal is absent. Genital pore is median.
8. Development is intra-uterine i.e. viviparous.
9. A remarkable feature is that a second embryo is produced within the first one, a third within the second and a fourth within the third. The first embryo, on completing its development passes out, still enclosing the other embryos, and attaches directly to a host fish.

GIGANTOCOTYLE (PARAMPHISTOMUM)

Phylum	-	**Platyhelminthes**
Class	-	**Trematoda**
Order	-	**Digenea**
Family	-	**Paramphistomidae**
Genus	-	***Gigantocotyle***

1. It is an endoparasite in the rumen of cattles and bile-duct of sheep, goat etc.
2. Body is dorsoventrally flatenned and measuring 12-14 mm in length.
3. Mouth lies at the anterior end and without an oral sucker.
4. Acetabulum is very prominant and lies at the posterior end.

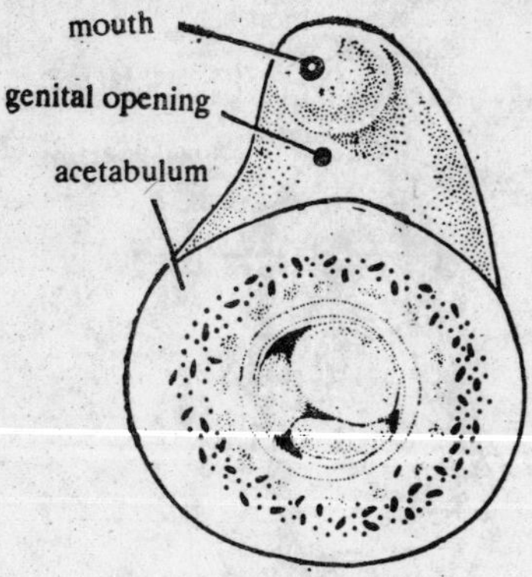

Fig. 1.14 *Gigantocotyle.*

5. Digestive system consists of mouth, highly muscular pharynx and bifurcated, unbranched intestine.
6. Parenchyma has a sort of delicate tubes forming much branched lymphatic system.
7. Testes are lobed and occur near the middle of the body.
8. Ovary is small, lobed and lies posterior to testes.
9. Vitellaria are scattered on the sides of animal.
10. Genital pore is just behind the bifurcation of intestine.
11. Uterus is slightly folded and eggs are very large.

GASTRODISCOIDES HOMINIS

Phylum	-	**Platyhelminthes**
Class	-	**Trematoda**
Order	-	**Digenea**
Family	-	**Paramphistomidae**
Genus	-	***Gastrodiscoides***
Species	-	***hominis***

1. It is an endoparasite in the colon and caecum of man.
2. It is found in India and Indo-China.
3. It measures 5-7 mm in length.
4. Body is differentiated into anterior and posterior part.
5. Terminal mouth is present an anterior part.

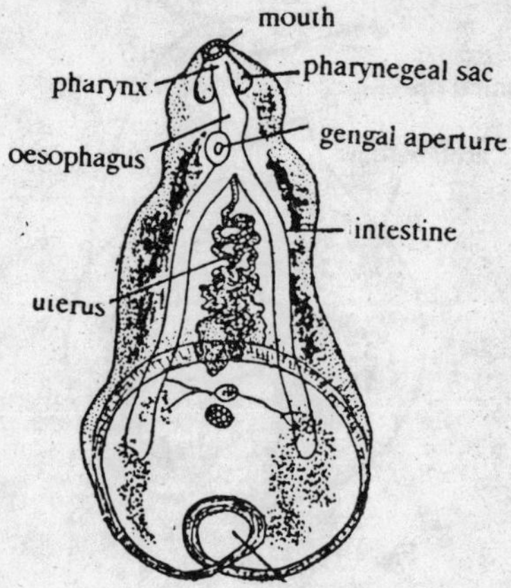

Fig. 1.15 *Gastrodiscoides*

6. Oral sucker is absent.
7. Posterior part bears a prominant acetabulum on ventral side.
8. Gonopore is in the anterior half of the body on ventral side.
9. Excretory pore lies on the dorsal side infront of acetabulum.
10. The secondary host is snail.
11. It causes inflammation and diarrhoea.

GASTROTHYLAX

Phylum	-	**Platyhelminthes**
Class	-	**Trematoda**
Order	-	**Digenea**
Family	-	**Paramphistomidae**
Genus	-	***Gastrothylax***

1. It is an endoparasite in the rumen of cattles.
2. It is elongated animal possessing a deep ventral pouch which extends upto the ventral sucker and it is supposed to be elongated genital chamber.
3. Digestive system includes mouth, pharynx, oesophegus and simple non-diverticulated intestine.

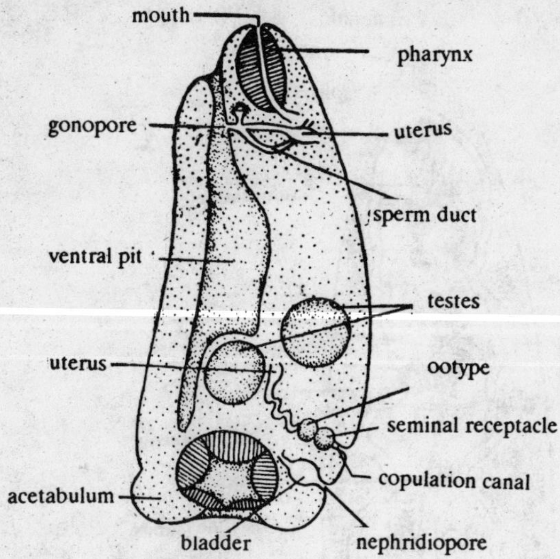

Fig. 1.16 *Gastrothylax*

4. Two rounded testes near the base of the pouch.
5. Female reproductive organs consists of single ovary, uterus, Ootype, vagina and seminal recepticle.
6. Single excretory pore is present.
7. Posterior end is divided with large acetabulum or adhesive sucker.

FASCIOLOPSIS BUSKI

Phylum	-	**Platyhelminthes**
Class	-	**Trematoda**
Order	-	**Digenea**
Family	-	**Fasciolidae**
Genus	-	***Fasciolopsis***
Species	-	***buski***

1. It is a large intestinal fluke of man, dogs, pigs etc.
2. It is found in India and China.
3. It measures about 70 mm in length and 20 mm in width.
4. Oral sucker and acetabulum both are well developed.

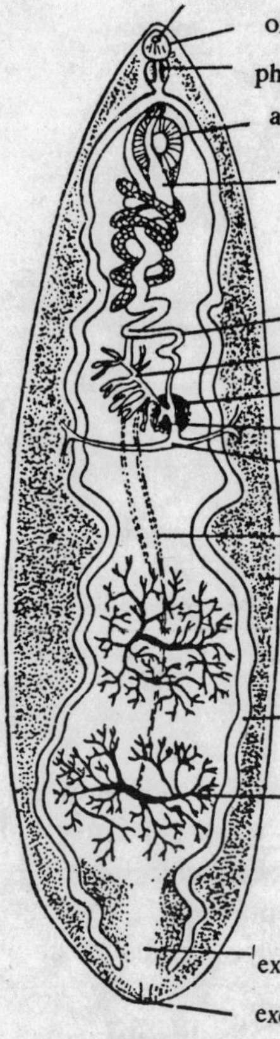

Fig. 1.17 Fasciolopsis buski.

5. Mouth surrounded by oral sucker leads into small pharynx. Oesophegus is short and leads into non-diverticulated caeca.
6. Gonads are present in posterior region of body.
7. Intermediate host is snail.
8. Cercaria larval encyst on tubers of water caltrop or singhara. The singhara is eaten raw and thus infection occurs.
9. It causes inflammation and hemorrhage of the intestine.

PARAGONIMUS WESTERMANNI

Phylum - Platyhelminthes
Class - Trematoda
Order - Digenea
Genus - *Paragonimus*
Species - *westermanni*

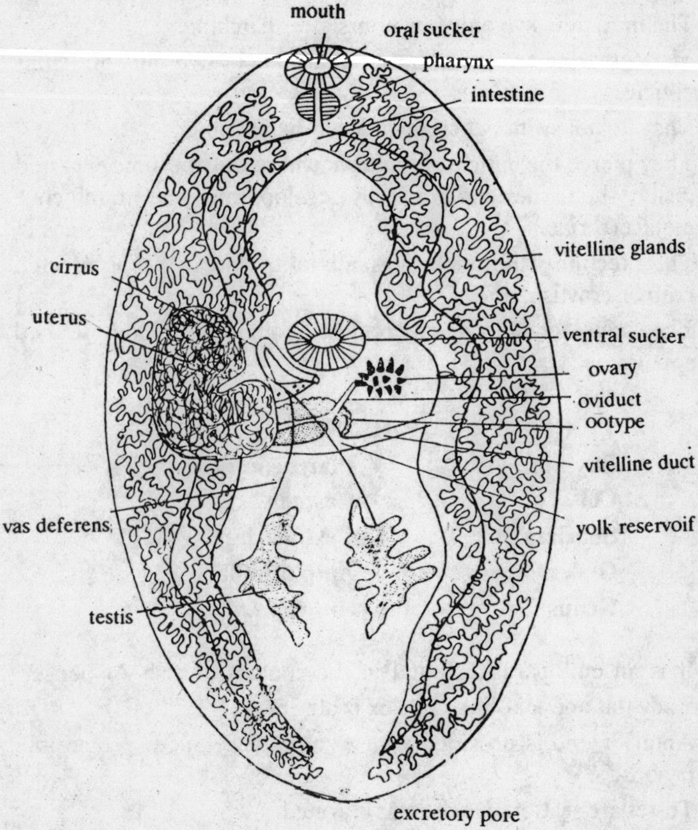

Fig. 1.18 *Paragonius.*

1. *Paragonimus westermanni* or Lung fluke is parasite in the lungs of mammal including man.

2. The body of the fluke is oval with well developed oral sucker surrounding the mouth and a ventral sucker in the centre of the body.
3. The internal organs are very much like that of Fasciola hepatica.
4. The cyst like pockets are formed around the parasite which rupture and liberate the eggs into bronchial tubes to be executed with sputum.
5. The miracidia live only few hours after hatching.
6. The cercariae are long, have small knob like tail and spiny cuticle.
7. They do not swim but creep in leech-like manner.
8. They pierce the cuticle of crayfish where they become encysted inside the tissues and gradually develop into mature infective metacercariae.
9. The infection of the new host usually takes place by eating infected crab or crayfish.
10. They cause or help in tuberculosis, cough and blood stained sputum.

AMPHILINA

Phylum	-	**Platyhelminthes**
Class	-	**Cestoda**
Subclass	-	**Cestodaria**
Order	-	**Amphilinida**
Genus	-	***Amphilina***

1. It is an endoparasite found in the coelom of fish, Acipenser.
2. Body flat and leaf-like. Scolex is absent.
3. Anterior end is provided with weakly developed protrusible proboscis.
4. Testes are scattered and cirrus is armed.
5. Vagina lies behind the ovary and opens on the left side. Uterus is coiled.
6. The intermediate host is a crustacean, Gammarus, which is eaten by Acipenser.

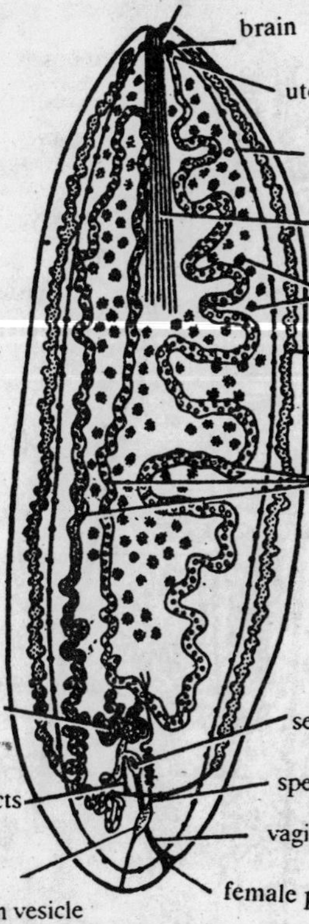

Fig. 1.19 *Amphilina.*

GYROCOTYLE

Phylum	-	**Platyhelminthes**
Class	-	**Cestoda**
Subclass	-	**Cestodaria**
Order	-	**Gyrocotylidea**
Genus	-	***Gyrocotyle***

1. These are endoparasite in the intestine of chimaeroid fishes.
2. Body is elongated and dorsoventrally flatenned.
3. Anterior end bears a large opening that leads into a highly muscular protrusible mass, the proboscis.

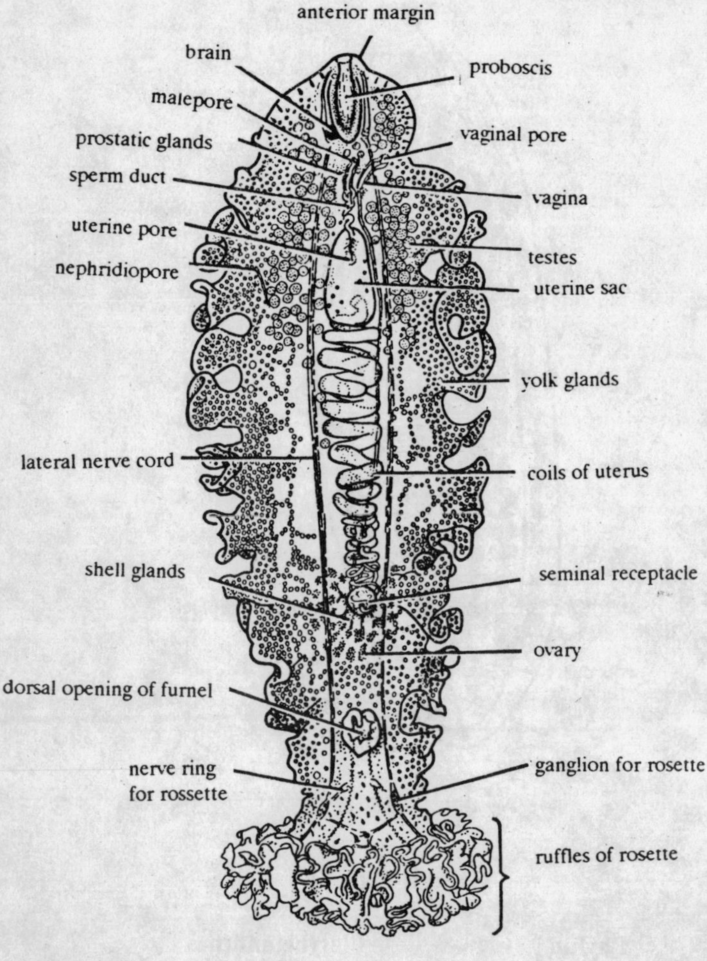

Fig. 1.20 *Gyrocotyle*

4. Posterior end bears a rosette-shaped adhesive organ, surrounding a funnel shaped depression.
5. The margins of body are generally ruffled.
6. Bisexual Male reproductive system comprise scattered testes, sperm duct and penis papilla. Female reproductive system comprises ovary, oviduct and uterus.
7. Follicles of yolk gland scattered throughout.

TETRARHYNCHUS

Phylum	-	**Platyhelminthes**
Class	-	**Cestoda**
Subclass	-	**Eucestoda**
Order	-	**Trypanorhyncha**
Genus	-	***Tetrarhynchus***

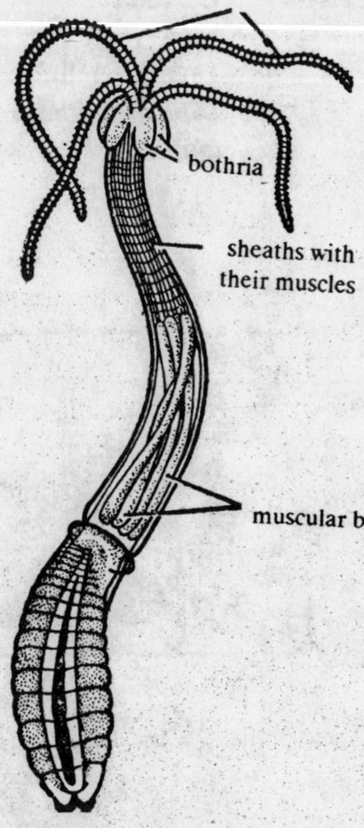

Fig. 1.21 *Tetrarhynchus.*

1. It is an endoparasite in the intestine of elasmobranch fishes.
2. Body is long divisible into scolex and proglottids.
3. Scolex is long and differentiated into a long proximal part containing the proboscis apparatus and a distal part bearing the bothria or sucker.

4. Scolex bears four suckers each with a eversible proboscide armed with spines. The proboscides are retractile in muscular sheaths which end in muscular bulbs.
5. Each proglottid contains a complete set of reproductive organs.

DIPHYLLOBOTHRIUM (DIBOTHRIOCEPHALUS)

Phylum	-	**Platyhelminthes**
Class	-	**Cestoda**
Subclass	-	**Eucestoda**
Order	-	**Pseudophyllidea**
Genus	-	***Diphyllobothrium***
Species	-	***latum***

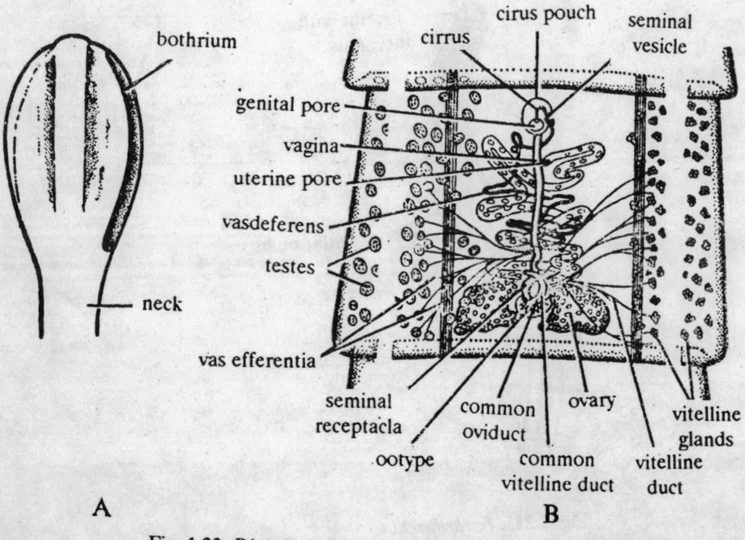

Fig. 1.22 *Diphyllobothrium* A--Scolex; B--A mature proglottid.

1. It is found in the intestine of man. It is world wide in distribution.
2. It is the largest and most pathogenic cestoda of human beings, measuring 20 metres in length.
3. Body is differentiated into scolex, neck and proglottids. The number of proglottids is 3000-4000.
4. Scolex is fusiform and bears two slit-like suckers or bothria.

5. Mature proglottid is broader than long.
6. Each proglottid is hermaphrodite and contains complete sets of male and female reproductive organs.
7. Testes are numerous. Vasa efferentia unite to form coiled sperm duct.
8. Ovary is trilobed. Vagina runs posteriorly forming seminal vesicle. Uterus very much twisted.
9. Gonopores ventral.
10. Life-cycle involves two intermediate hosts, one is *cyclops* (a copipod) and other a fish.
11. It causes bothriocephalus anaemia, erythropenia and haemorrhage.

MONIEZIA

Phylum	-	**Platyhelminthes**
Class	-	**Cestoda**
Subclass	-	**Eucestoda**
Order	-	**Taenioidea**
Genus	-	***Moniezia***
Species	-	***expansa***

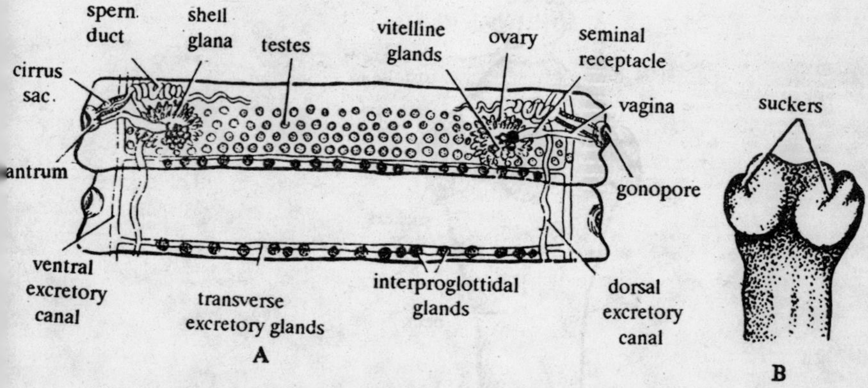

Fig. 1.23 *Moniezia*. A–Two proglottids. B--Scolex.

1. It is found in sheep, cattles and other ruminants.
2. Body is divisible into scolex, neck and proglottids.
3. The scolex is small with prominant suckers. The rostellum and hooks are absent.

4. Each proglottid is broader than long and contains a double set of reproductive organs.
5. Genital ducts are dorsal to osmoretgulatory canals.
6. Testes are numerous scattered throughout the centre of proglottid. Cirrus in cirrus pouch.
7. Ovaries are in the form of an open fan. Uteri form networks.
8. On one side vagina is dorsal to cirrus pouch and on other ventral.
9. The intermediate host is mite (*Galumna*).
10. It is not so pathogenic but sometimes may cause diarrhoea or anaemia.

HYMENOLEPIS NANA

Phylum	-	**Platyhelminthes**
Class	-	**Cestoda**
Subclass	-	**Eucestoda**
Order	-	**Taenoidea**
Genus	-	***Hymenolepis***
Species	-	***nana***

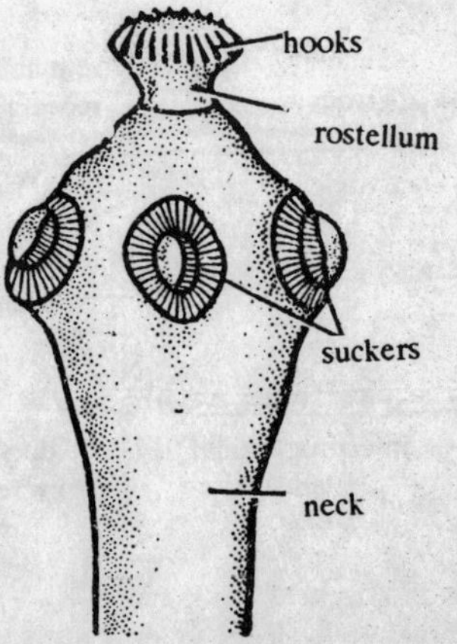

Fig. 1.24 Scolex of *Hymenolepis.*

1. It is commonly called the dwarf tapeworm of man. It lives in the intestine.
2. It is world-wide in distribution.

3. It measures 25-40 mm in lengths and has about 100-200 proglottids.
4. The rostellum is well-developed and retractile. It bears a single circlets of 20-30 hooks.
5. Neck is long and slender.
6. There are three testes with varying arrangement. Cirrus pouch is large with external and internal seminal vesicle and with an accessory sac.
7. Life-history is simple without any intermediate host.
8. The inner membrane of the hexacanth bears long, wavy filaments.
9. Larvae develop in the intestinal villi while the adult lives in the intestine.
10. It causes abdominal pain and diarrhoea.

DIPYLIDIUM CANINUM

Phylum	-	**Platyhelminthes**
Class	-	**Cestoda**
Subclass	-	**Eucestoda**
Order	-	**Taenioidea**
Genus	-	***Dipylidium***
Species	-	***caninum***

1. It is commonly called dog-tapeworm.
2. It has world wide distribution.
3. It lives in the intestine.
4. It measures 15-40 cm in lengths and has under 200 proglottids.
5. The scolex is somewhat triangular and bears four deeply cupped suckers.
6. Rostellum is retractile into a rostellar sac in scolex. It bears about 60 hooks in 3-7 rings.
7. Mature proglottids are long and barrel shaped having a double set of male and female sex organs. The genital pores at the middle of both lateral margins.
8. Numerous testes are present before or behind the female genitalia.
9. The uterus first develops as a network of canales, but latter breaks up into a large number of uterine capsules, each enclosing 3-30 embryos.

10. Intermediate host is dog-fleas, *ctenocephalus canis* and *pulex irritans*.

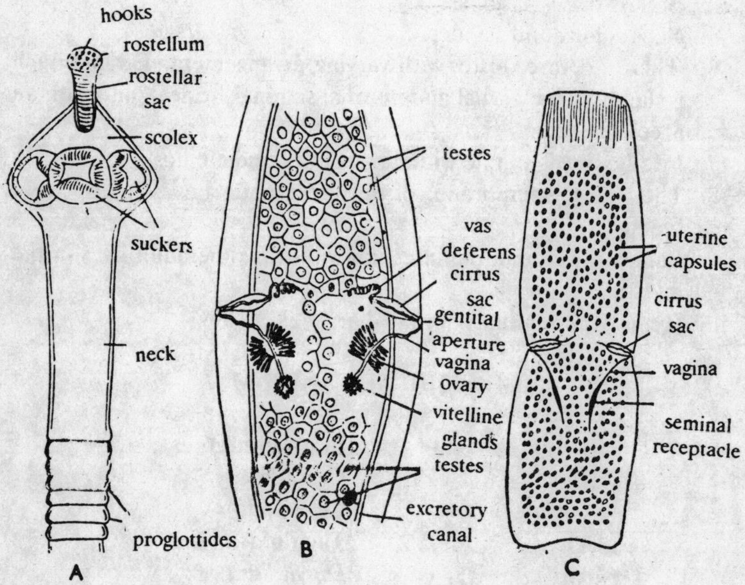

Fig. 1.25 *Dipylidium*. A–Anterior part. B–Mature proglottis. C–Ripe progloltis.

TAENIA SAGINATA

Phylum	-	**Platyhelminthes**
Class	-	**Cestoda**
Subclass	-	**Eucestoda**
Order	-	**Taenoidea**
Genus	-	***Taenia***
Species	-	***saginata***

1. It is commonly called the beef tapeworm. It lives in the intestine of man where beef is eaten.
2. The body is dorso-ventrally flatenned and much larger than taenia solium. It measures 15-20 feet in length.
3. Body is divisible into scolex, neck and proglottids.
4. Scolex bears four large suckers, adhesion and devoid of rostellar hooks.

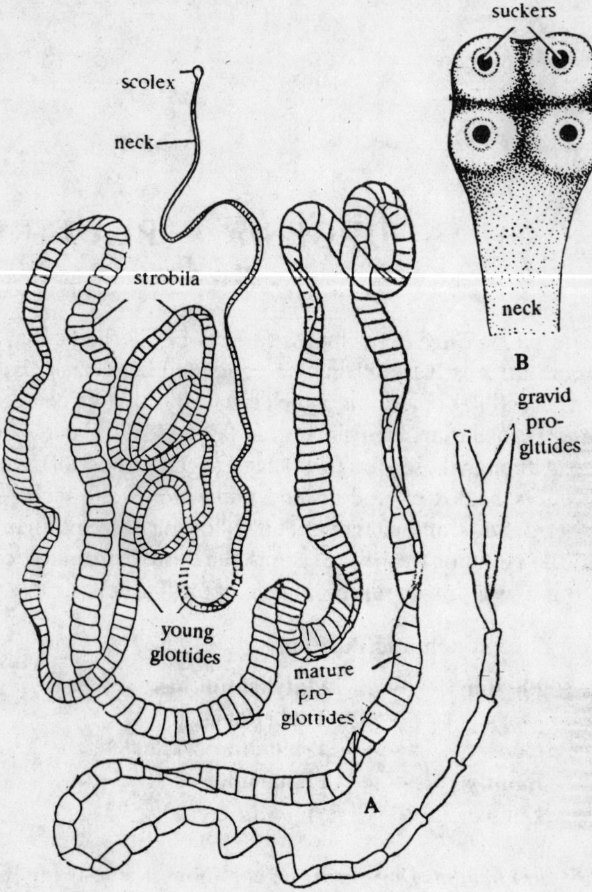

Fig. 1.26 *Taenia saginata*

5. About 1000 proglottids are present.
6. Testes are about 300-400. Ovary in the form of two lobes. Ootype is present surrounded by Mehli's gland.
7. Life-history is similar to *T. solium*.
8. Intermediate hosts are cattles.

2

DUGESIA (PLANARIA)

Planaria or *Dugesia* belongs to class Turbellaria. The Turbellarians are mostly free-living, unsegmented flatworms that are clothed with a cellular epidermis, which is usually ciliated, sometimes only in parts. Turbellarians are primarily aquatic, and the great majority are marine. Although these are a few pelagic species, most of them are bottom dwellers that live in mud or sand, under stones and shells, or on sea weeds, *Degusia* is among them. The following description mostly belongs to the common fresh water genus, the *Degusia* formarly called *Planaria*. It is a well-known representative of the Turbellaria.

Systematic Position

Phylum	-	**Platyhelminthes**
Class	-	**Turbellaria**
Order	-	**Tricladida**
Family	-	**Planariidae**
Genus	-	***Dugesia***

Habits and habitat. Dugesia is the commonest free-living platyhelminthes. They are usually found on the vegetation or under the stones in ponds, lakes, marshes, rivers and small streams. They generally inhabit pure, cool and clear water. Some species remain in currents, while others like to remain in quiet water. When active, they glide over the surface but when at rest they crawl down to the bottom or under other smooth objects, where light is not bright. *Dugesia* is found under the stone generally attached in great numbers, if pieces of meat are dropped in water near the bank.

It starved they become smaller, but if well-fed they lay cocoons containing eggs which hatch to give young. They are carnivorous and extremely cannibalistic.

MORPHOLOGY

Shape, Size and Colour

These are small, thin, dorsoventrally flatenned, bilaterally symmetrical animals measuring about 12 mm long. Colouration is due to the dissolved pigments present beneath the epidermis. The colour may be dark yellow or olive, brown or blackish brown and may have stripes of white, orange or yellow colour. The colouration is light or whitish on ventral side.

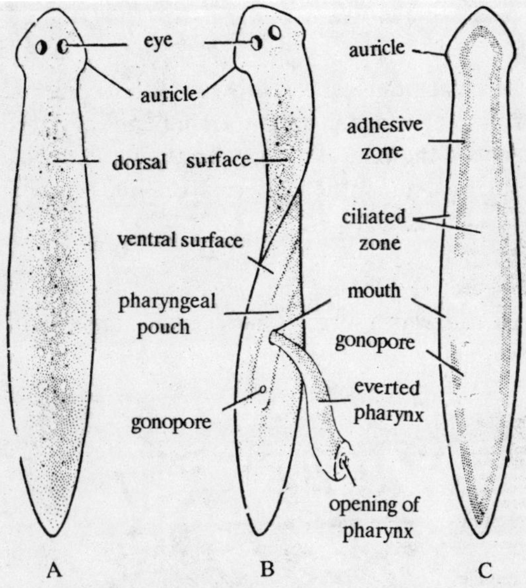

Fig. *Dugesia.* External features. A = Dorsal view; B = Lateral view; C = Ventral View

Structure

There are two ends in the body. Anterior end is blunt, while the posterior end is tapering. Anterior end is almost square or triangular with its apex pointing forward. This portion of the body is known as head, because with it animal moves forward. There are two lateral projections, called '*auricles or ears*' at the base of the head. But these ears are not sensory organs. On the dorsal side of the head, near the anterior end, there are two black *eye-spots*. Behind the auricles the body is somewhat constricted to form a so called *neck*. On the ventral side of body, near its middle, is found the mouth through which *proboscis* or *pharynx* protrudes. Thus, in Dugesia mouth is not situated on the head. The muscular walls of the pharynx serve for capturing the food. Behind the mouth in mature worms is situated a small *genital pore.* There are also excretory openings, the *nephridiopores,* on the dorsal surface situated laterally, but they are very minute and are, therefore, invisible. The ventral surface of the body is covered with cilia, which help in locomotion.

Body-wall

The body of a planarian is covered by a ciliated epidermis. The epidermal cell on its free surface also show a number of microvilli. Cilia are more prominent on the ventral surface. The epidermal cells rest on a basement membrane. Among the epidermal cells the following other types of cell are also present-

(i) *Gland cell* secreting adhesive substances.
(ii) *Rhabditogen cell* which secretes and contains rhabdites.
(iii) *Sensory cell.*

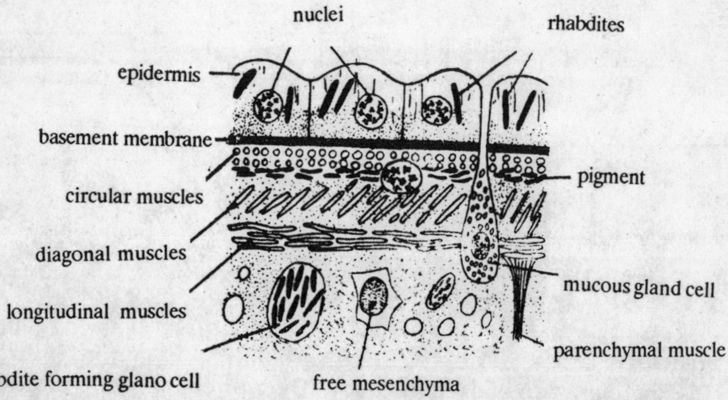

Fig. 2.2 *Dugesia*. T.S. of body wall to show ciliated epidermis and muscle layers.

Rhabdites are arranged at right angles to the body surface. They are hyaline elliptical structures secreted by rhabditogen cells. When they are expelled out they swell on contact with water forming a sticky froth around the body. This may be protective, help in gliding or even in capture of food. Some workers like *Reisinger* and *Kelbetz* (1964) suggest that the rhabdites and nematocysts are evolutionarily related.

Below the epidermis are granules and rods of pigments. The gland cells are unicellular, some occur in the epidermis but most of them are in the mesenchyme. They have long necks opening on the surface and secrete mucous.

Below the epidermis is a thin elastic basement membrane at which the epidermal cells rest. It maintains the general form of the body and provides surface for the attachment of underlying muscles.

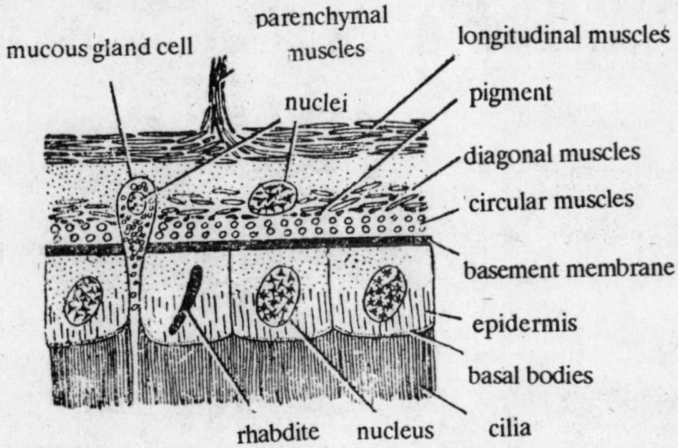

Fig. 2.3 *Dugesia*. V.I.S. Ventral body wall.

The muscle layer consists of outermost this sheet of *circular muscle fibres*, middle broad *diagonal muscles* and innermost is the longitudinal muscles. On the ventral body wall the *longitudinal muscles* are more developed as this surface is important in the locomotion of the animal. Some muscle fibres run dorsoventrally in mesenchyme-*Mesenchyme*. The bulk of body-wall and the interior is filled with mesenchyme, which consists of connective tissue (net-like syncytium containing nuclei), free wandering cells and fluid filled spaces. The wandering cells multiply by mitosis and differentiated into other types of cells. These cells help in regeneration in the *Dugesia*. Parenchyma or mesenchyma helps in transport of food, assimilation and in the excretion of waste products.

PHYSIOLOGY

Locomotion.

Dugesia is aquatic but it does not swims. The animal exhibits two types of movements.

1. Gliding movement
2. Crawling movement

1. *Gliding movement*- In gliding movement the cilia present on the ventral surface beat against a track of mucus secreted by the mucous glands. As a result of ciliary action, the animal is propelled forward in a gliding manner. The wave of ciliary action proceeds from the anterior to the posterior end and the head of the animal is slightly elevated.

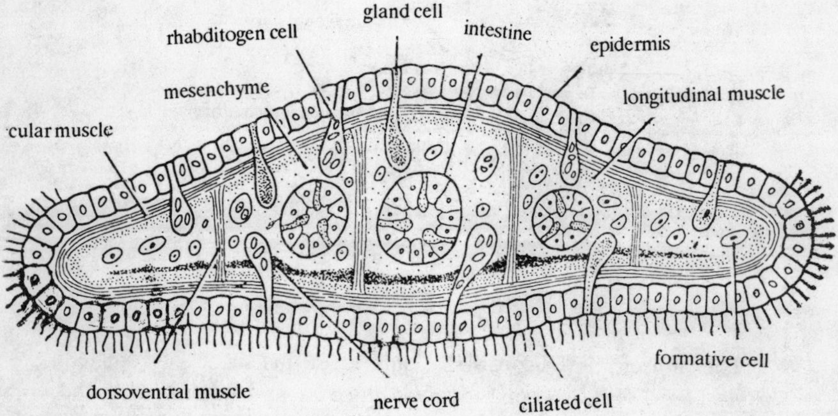

Fig. 2.4 *Dugesia.* T.S. body showing the disposition of muscles, nerve cord etc.

2. *Crawling movement* - It is brought about by the contraction of muscles. The wave of contraction extends from anterior to the posterior end, contraction of circular and dorsoventral muscle elongate the body, the anterior is held fixed to the substratum by mucous and the posterior end is then pulled forwards by contraction of longitudinal muscles.

Sometimes transverse waves of contraction sweep the length of the body causing its undulations in the vertical plane and enabling the planaria to crawl forward. Differential action of local muscle groups produces turning or twisting movements.

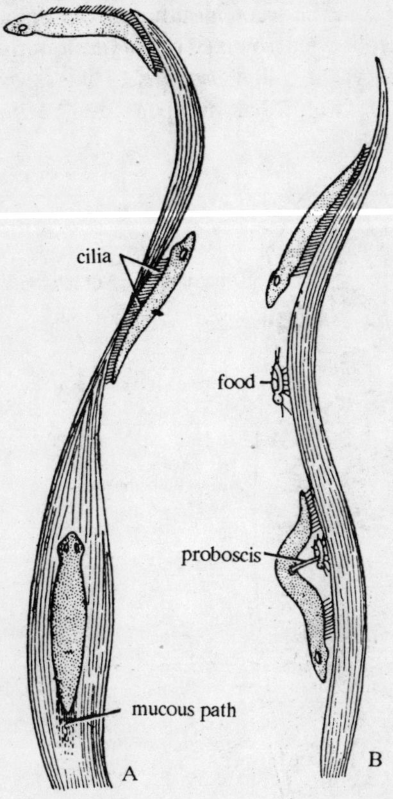

Fig. 2.5 *Dugesia*. A = Locumotion, B = Feeding.

Digestive System

1. *Alimentary canal*. The alimentary canal is incomplete i.e. there is only one opening for ingestion and egestion. The alimentary canal comprises : mouth, plarynx and intestine.

Mouth. The mouth is an oval or rounded aperture situated mid-ventrally a little behind the middle of the body. It is surrounded by radial and circular muscle fibres from the subepidermal musculature. The muscle fibres regulate the opening or closing of mouth. It leads into a large cylindrical elongated chamber. The *pharyngeal pouch*.

Pharynx. The pharynx is a tubular organ situated within the pharyngeal pouch projects posteriorly. The pouch is bounded by a muscular sheath, the *pharyngeal sheath*. A space sometimes called buccal cavity is present between mouth and the opening of pharynx. The free end of the pharynx projected out of the mouth during feeding. The projecting pharynx is called *proboscis.* This is brought about by the action of muscular wall. Pharynx opens into the *intestine* by way of a short *oesophagus.*

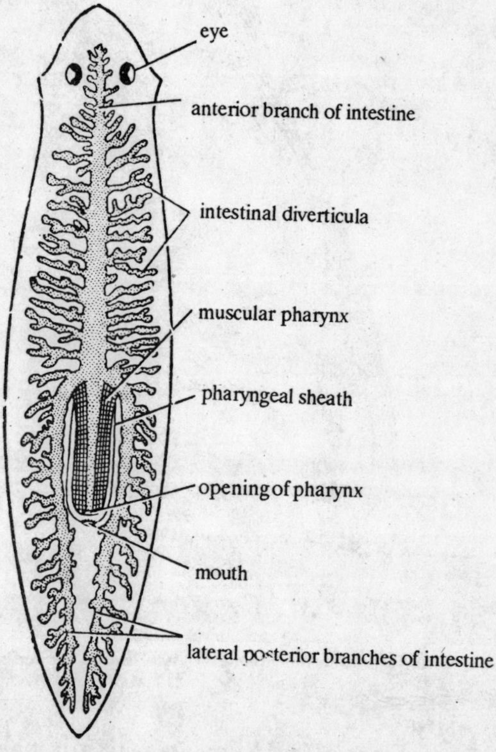

Fig. 2.6 *Dugesia.* Alimentary canal.

Intestine. The intestine immediately divides into three branches. The median branch extends forwards upto the head while the other two lateral branches extend backwards upto the posterior end, one on either side of pharyngeal cavity. All the three branches give off numerous branching and blind diverticulal which end into the mesenctryme. The much branching intestine is a means of increasing the surface area for digestion, absorption and distribution of food.

Histololgy of alimentary canal. The pharyngeal wall shows a complicated histological structure. It consists of nine layers from the surface to the lumen : epithelial cells, longitudinal muscle layer, circular muscle layer, outer gland cells, nerve plexuses, inner gland cells, longitudinal muscle layer, circular muscle layer and an endodermal epithelial lining.

The intestine is simple and thin consists of a single-layered epithelium, the gastrodermis. The gastrodermal cells are of two types phagocytic cells and granular cells.

Food and feeding. Normally *Dugesia* is *holozoic* and canivorous subsisting for the most part on crustaceans, insect larvae and small worms. During feeding, it glides through the action of cilia and the adhesive glands secrete adhesive secretion. When prey comes in contact with the secretion it becomes firmly adherent. *Dugesia* then folds the anterior end of its body over the prey and thus immobilizes it. The pharynx is then produced out through the mouth and through its peristaltic action sucks up small particles of the prey into its intestine.

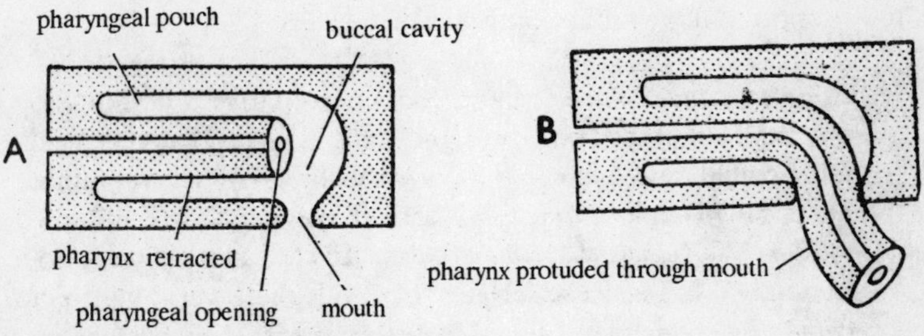

Fig. 2.7 *Dugesia.* Diagrammatic representation of : A = Retracted pharynx. B = Protruded pharynx.

Digestion. Digestion is initially extracellular. Food is broken down by pharyngeal proteolytic enzymes and endopeptidases secreted by the gland cells lining the intestinal wall. The broken pieces of partially digested food are then digested by intracellular method. The phagocytic cells of the intestine engulf these particles which are digested by intracellular endopeptidases inside cytoplasmic vacuoles in acidic medium. After about 8 hours the vacuolar content becomes alkaline and digestion in this medium is completed by exopeptidases, lipases and carbohydrases. The excess of food is stored as fat and proteins in the intestinal cells and mesenchyma.

Freshwater planarians are able to withstand prolonged experimental starvation. In extreme cases, they utilize part of enteron and all the mesenchyma and reproductive system. In fact, the body volume may be reduced to as little as 1/300 of the original.

Egestion. Since there is no anus, the egestion of undigested food material takes place through mouth. It is brought about by the contraction of the muscles of body-wall.

Respiration. Mode of respiration is aerobic but there are no special organs for this. Oxygen from the surrounding water is directly into the body and carbon dioxide is diffused out through the entire body surface. Flatenned body is an admirable adaptation for this type of respiration

Excretory system. The excretory system or *protonephridial system* of *Dugesia* consists of a pair of *longitudinal excretory canals* and numerous tiny enlargements known as flame cells.

(i) *Excretory canals.* A pair of longitudinal excretory canals running on each side of the body throughout the length. They are highly inter-coiled and open to the dorsal surface by several minute poses called nephridiopores. Anteriorly two canals are united by a *transverse vessel*, in front of eyes. Each canal divides into a number of branches which further divides into very fine blind *capillaries* which bear flagella at the inner end (*Wilson* and *Webster*, 1974). The terminal structure is composed of flame cells or cryptocytes.

(ii) *Flame cells.* Each *flame cell* represents the unit of excretory system. It is a large, tubular, protoplasmic body. It is produced into numerous protoplasmic processes reaching the mesenchynme. Its terminal part is swollen into a bulb and contains a conspicuous nucleus. The flame cell has an intracellular space which is continued into capillaries. The space encloses a bunch of cilia which vibrate giving the appearance of flickering candle flame.

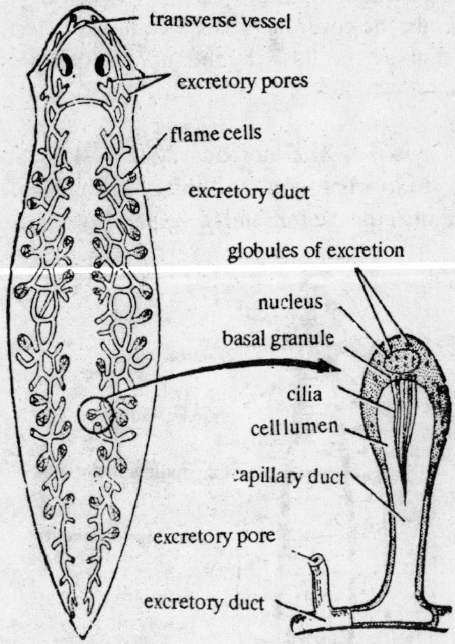

Fig. 2.8 *Dugesia.* Excretory system.

Besides the flame cells, certain glandular cells called the *arthrocytes* or *paranephroytes* occur in close contact with the excretory canals and also help removal of excretory products.

Physiology of excretion. Excretory substance is collected from the mesenchyme and is transferred into the cavities of flame cells. The beating of cilia of flame cells causes hydrostatic pressure by which the fluid waste passes into the longitudinal trunks and goes out of nephridiopores.

According to *Storer, Wolcott, Guyer, Hickman* etc. the flame cells simply help in excretion of nitrogenous wastes. But according to some workers like *Mckanna,* 1968; *Prusch,* 1976 etc. the excretory system in *Dugesia* is largely Osmoregulatory bailing out excess of water. As osmoregulatory organs, the protonephridia might function in the following manner. The beating of cilia of flame cells would produce a negative pressure within the end of the tubule, causing filteration of fluid across the membrane covering the fenestrations. Selective reabsorption of ions, especially K+ and Cl-, by the tubule would yield a hyposmotic fluid.

Nervous system. The nervous system shows a much better development than what we found in the coelenterates and can be differentiated into the *central* and *peripheral systems.*

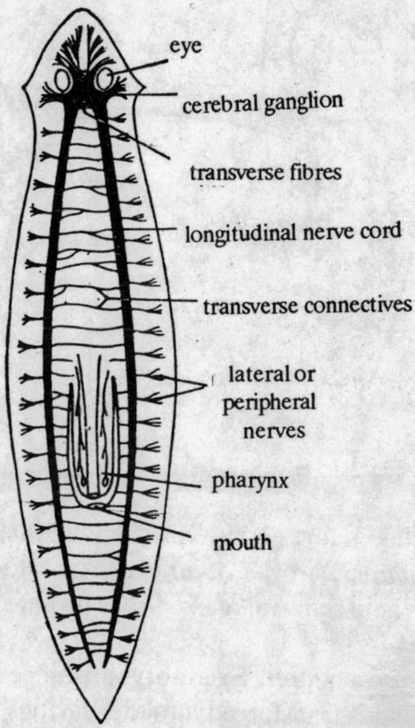

Fig. 2.9 *Dugesia.* Nervous system.

Just below the epidermis and behind the eyespots, there is a bilobed *brain* or *cerebral ganglia* in the form of an inverted 'V'. Both the ganglia are connected by several transverse fibres. From the brain, several thin nerves are given off anteriorly and to the eyes. From the brain two *lateral longitudinal nerve cords* extend posteriorly along the ventral side. The two nerve cords are connected together by *transverse connectives.*

The peripheral system is present in the form of a *sub-epidermal plexus* (just below the epidermis) and a *sub-muscular plexus* below the muscle layers in the mesenchyme.

The nerve-cells or neurons may be unipolar, bipolar or multipolar. All the neurons possessing neurites are similar. These neurites cannot be differentiated into axons and dendrons.

Sense organs. Dugesia possesses special sensory organs, called receptons. These are photoreceptors or eyes, auricular organs, tangoreceptors and rheoreceptors.

(i) *Photoreceptors or eyes*. The eyes are two dark spots on the dorsal surface of the anterior end, one on either side of the median line. Each eye has the form of a cup whose opening faces anterio-posteriorly.

Structure of each *eye* is very simple. It is made up of highly pigmented *retina.* Inside the cup are situated numerous nerve-cells, the *fibrillae* of which are in contact with the retina and their opposite ends are continued into an optic nerve, which is joined with *brain*. The nerve cells enter the cup through the opening. Ectoderm over the eye is not pigmented. *Lens* is absent in eye.

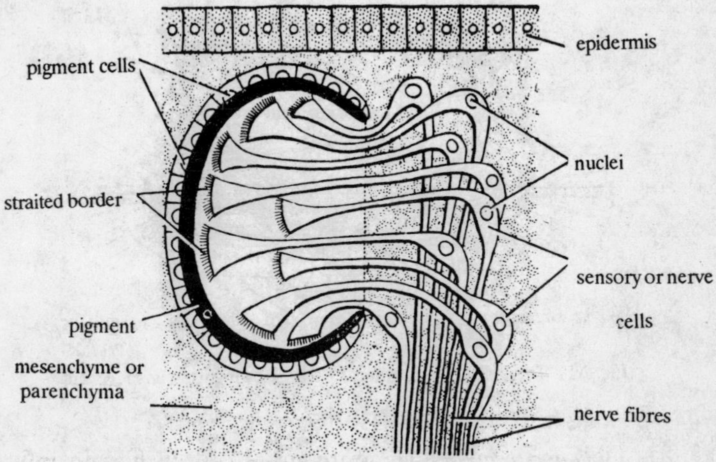

Fig. 2.10 *Dugesia.* V.S. of eye.

Eye are capable of a crude discrimination of the direction of light. The pigment cup serves as a shield and light can enter only through its opening to stimulate the photosensitive expanded ends of retinal cells, thus the animal can detect the direction light. The animal is negatively phototactic and is most active at night. It the eyes are removed it can still react to light. However the reaction is much slower. This shows that it possesses some light sensitive cells over the general body surface.

(ii) *Ciliated pits*. Besides eyes, there are *ciliated pits* which are also sense-organs. They are situated on either side of the head, having special sensory cells with long cilia. These ciliated pits are connected with nervous network. Tentacles, which are found in some turbellaria with long cilia, are also supposed to be organs of chemical sense, which help in finding food.

(iii) *Auricular organs.* These are sensitive to chemical and are chaemoreceptors. These are in the form of a pair of whitish ciliated groores, lying one on either side of the head near the base of the auricles. If the auricles are removed the planarian cannot locate food.

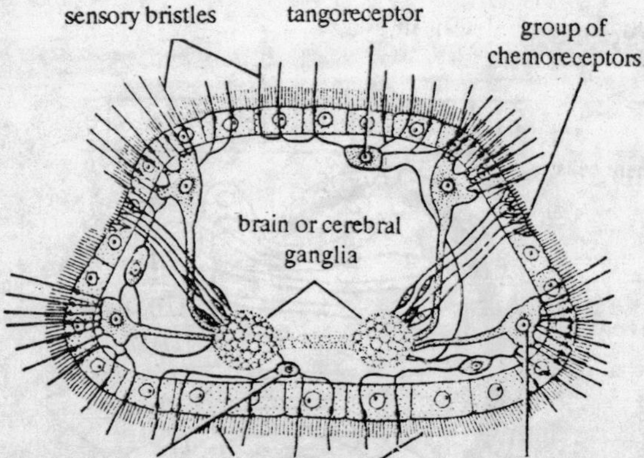

Fig. 2.11 *Dugesia.* Digrammatic T.S. Showing various receptors.

(iv) *Tango receptors*. These are sensitive to touch temperature etc. A tangoreceptor is a small cell with fine sensory bristles that project over the surface except one that posses to the brain. They are particularly concentrated on the ventral surface around the mouth, auricles and the body margins. Thus ventral surface is positively thigmotactile while the dorsal surface is negatively thigmotactic.

(v) *Rheoreceptor.* These are sensitive to water current. They are of general occurrence and their sensory processes or bristles project much beyond the level of cilia.

Reproductive system

Dugesia reproduces with asexually as well as sexually

(1) Asexual Reproduction

Asexual reproduction occur by the transverse binary fission. The plane of fission usually forms posterior to the pharynx. The posterior part attaches to substratum by mucus, the anterior part stretches forward until the animal snap into two. Each half regenerates the wanting parts resulting in the formation of two complete planarians. Some planarians

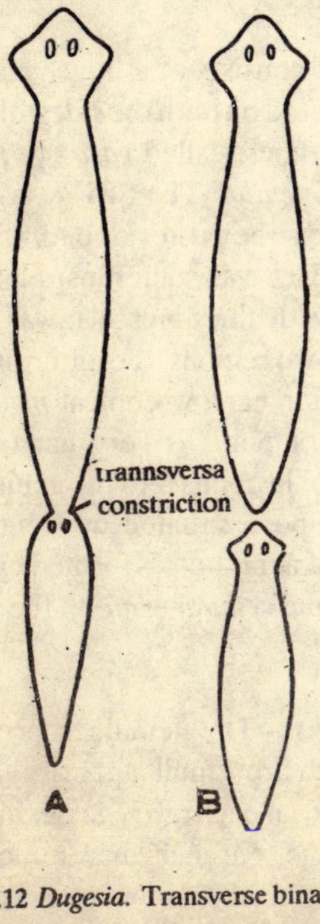

Fig. 2.12 *Dugesia.* Transverse binary fission.

fragment into two or more picccs each growing into a complete individual. Asexual reproduction is restricted to some species only. Some species reproduce asexually in one season, and may develop sex organs in the other, but these are some species that reproduce exclusively asexually and rarely develop sex organs.

(2) Sexual Reproduction

The reproductive organs are more complicated in these animals. The gonads i.e. testes and or aries are derived from the meroderm and there is a system of tubules and chambers in which fertilization occurs.

The animal reproduces in early summer and the gonads are developed temporarily in the breeding season. The *Dugesia* is bisexual or hermaphrodite but self-fertilization does not take place and the cross fertilization is the rule. After breeding season the gonads degenerate and disappear and the worm reproduces asexually.

Male reproductive system consists of a large number of small spherical testes, which are situated on both the sides of the body. Each testis is connected with minute tubes, called *vasa effemtia.* There form two large ducts called *vasa deferentia.* The two vasa deferentia run parallel to one another towards the posterior end, where these bend mesially towards the penis. Each vas deferens dilated to form the seminal or spermlducal vesicle with thick muscular wall. The sperms are stored here. The penis shows two regions, the anterior large spherical *penis bulb* and the posterior narrow conical *penis papilla* with muscular wall. The lumen of penis is very narrow and called ejaculatory duct. The penis papilla opens into the genital atrium which terminates in the common gonopore, situated midventrally behind the opening of month. The lumen of penis bulb is formed by *bulbar cavities.* The bulb contains numerous uni *cellular glands,* the prostate glands which open into it lumen.

Female reproductive organs - The female reproductive organs consist of a pair *ovaries,* which are small apherical bodies situated laterally just behind the head. From each ovary arises an oviduct which extends posteriorly through the parenchyma. Immediately after its emergence each oviduct presents a dilation, which is known as *seminal receptacle*. The two oviducts unite to form a common vagina which

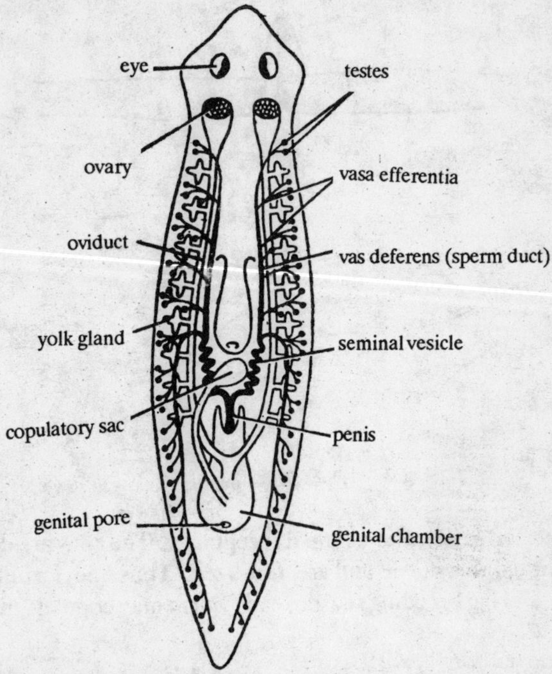

Fig. 2.13 *Dugisia* Reproductive system.

opens into the *genital artium*. The oviducts along their whole length are surrounded by numerous vitelline glands which open into oviducts by short *vitelline ducts*. Because of the close association of oviduct and vitelline glands, these are sometimes collectively referred to as the ovo-vitelline ducts. Opening into the genital chamber is found a muscular sac-like *copulatory sac* or *uterus* or *bursal canal*. The bursal canal opens into a large sac the *bursa conplatrix* which receives penis and temporarily stores the sperms received during copulation. It opens below into female genifal atrium.

Copulation

Although *Dugesia* is hermaphrodite, yet *cross-fertilization* is a rule. *Copulation* takes place between the two worms in which their posterior and ventral surface come together and the penis of each is inserted into the *genital atrium* of the other. In this way sperms are deposited in the *seminal receptacles* of the mating partner. The sperms are stored inside the copulatory sac for some time and then travel along

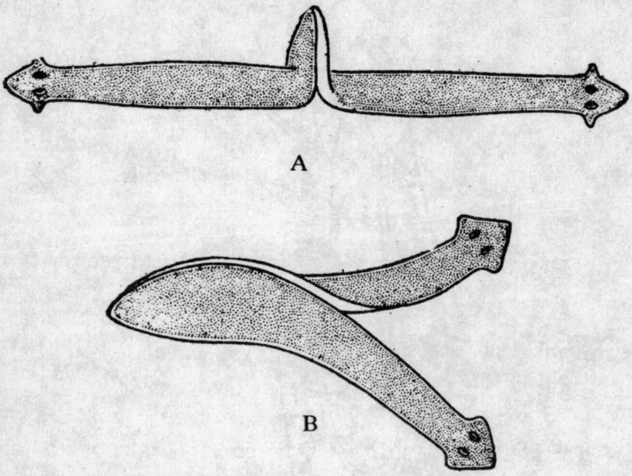
A

B

Fig. 2.14. *Dugesia.* Copulation

the oviducts to reach the seminal receptacles. The eggs are deposited in the seminal receptacle and are fertilized. Thus the fertilization is internal. During breeding season, each worm may copulate many.

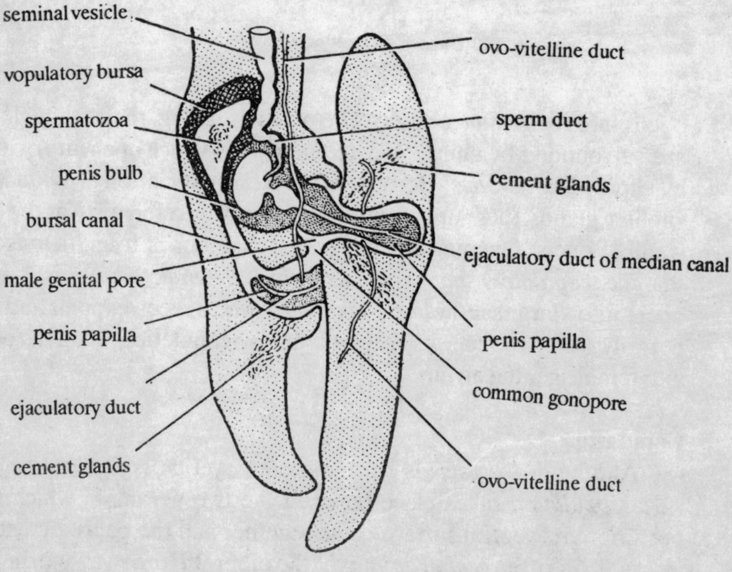

Fig. 2.15 *Dugesia.* A pair of copulating animals in section.

Cocoon formation.

The zygote thus formed, pass backwards through the ovovitelline ducts and reach the genital atrium of ovovitelline ducts and reach the genital atrium of female. In their way they are mingled with yolk cells which are discharged by yolk glands into the oviducts. The diposition of yolk outside the zygote is unique condition in platyhelminthes. Such eggs are called *ectolecithal*. When the eggs are reached in genital atrium 5-10 eggs along with yolk cells are enclosed in a protective, proteineous shell forming a cocoon or an egg-capsule. The capsule is coated by a sticky substance secreated by the cement glands that surround and open into the common genital atrium. The capsule is now extruded through the genital pore. As it is passing out, sticky coating is drawn out into a sticky stalk that fastens it to some object chiefly the under surface of stones. The capsules are laid in succession at intervals of a few day.

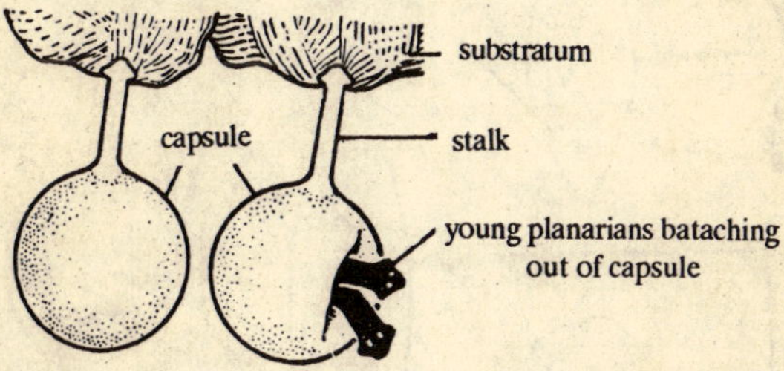

Fig. 2.16 *Dugeisa.* Young one haching from the cocoon.

Development

Development starts soon after the cocoon is laid and the hatching occurs in 2 to 3 weeks depending upon the temperature of water. The cleavage is *spiral* and *determinate* type. No larval stage is formed hence the development is *direct*. A young planaria or *juvenile* glides out of the capsule from each egg.

Regeneration.

Like *Hydra, Dugesia* shows extensive regeneration. The lost parts can be easily replaced. If the animal is cut transversely into two halves, the anterior half regenerates a new Tail, while the posterior half regenerates a new head. If the animal is cut into three pieces, the middle piece will regenerates head on the anterior side and a tail on posterior side. The regenerating piece retains the original *linear polarity* of the whole animal. Regeneration involves *epimorphosis* and *morphollaxis*. In epimorphosis the missing parts are formed while in morphollaxis, the original parts fit to function with the regenerated parts in the new individuals.

Power of regeneration is maximum near the head and minimum near the posterior end. this is associated with the metabolic rate. The metabolic rate is highest in the head and becomes progressively lower towards the tail. This variation in metabolic rate and regeneration power is called *axial gradient* (*Child*).

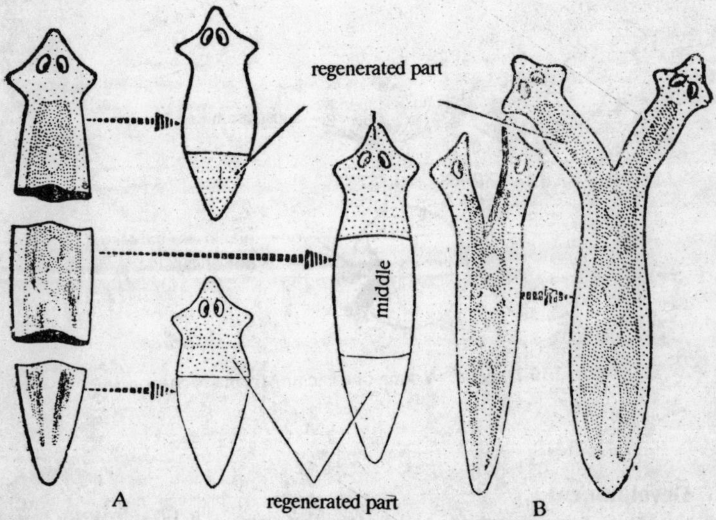

Fig. 2.17 *Dugesia* Regeneration. A = Transverse, B = Longitudinal

Regeneration is brought about by the free formative cells or *neoblasts* of the mesenchyme. These cells migrate to the cut surface and by repealed division produce new tissues.

Regeneration is always not perfect. It a sexually mature *Dugesia* is cut transversely between its pharynx and copulatory apparatus, its reproductive organs degnerate and fail to regenerate, but each help grows into asexual animal.

Dugesia shows a *lateral polarity.* If the anterior end is cut longitudinally, each part will grow into a complete head. Thus several headed monster results. The phenomenon is called *hetermorphosis.*

3

FASCIOLA HEPATICA

Fasciola belongs to class Trematoda. Most of the trematods living in other animals inhibit the digestive tract or its accessory tubes or cavities, and it seems probable that those species living outside gut are derived from gut dwelling forms. Probably the most important of these flukes, *Fasciola hepatica*, was the first digenetic fluke to have its life-cycle uncovered. *Thomas* (1883) described its life-cycle. *Reinhard* (1957) redescribed its life-cycle in an interesting way.

SYSTEMATIC POSITION

Phylum	-	**Platyhelminthes**
Class	-	**Trematoda**
Order	-	**Digenia**
Family	-	**Fascioliodae**
Genus	-	***Fasciola***
Species	-	***hepatica***

Geographical distribution

World wide distribution occurs in areas where sheep, cattle and goats are raised and there is a niche for lymnaied snails. In U.S.A., organisms found in the Gulf Coast, the North-West, Rocky Mountains, Florida etc. In U.S.A., it is endemic.

Habits and Habitat

Fasciola hepatica is derived from Greek and Latin words - (L., *fasciola* = small bandage ; Gr., *hepar*, = liver) is a *digenetic parasite*, i.e., its life history is completed in two hosts - a primary vertebrate host and a *secondary* or intermediate invertebrate host. Adult *F. hepatica* occurs in the liver and bile passages of the primary host which is usually sheep; but, at times, it may occur in some other vertebrates, like goat, horse, dog, ass, ox, deer, antelope, rabbit, elephant, monkey, etc. Humans are occasional host for *F. hepatica*, and infections are most often obtained through eating watercress (*Nasturtium officinale*) in salad. A single host may harbour as many as about 200 of these flukes in its liver, which may consequently cease to function normally. This effect is known as liver-rot. *F. hepatica* spends a part of its life history in an intermediate host, a freshwater snail which is either *Limnea truncatula* or some specific species of *Planorbis, Fossaria modicella, Pseudosuccinea* etc.

Shape, Size and Colouration

The body is large, soft and fleshy. It is dorsoventrally flatenned and leaf-like having somewhat oval appearence. The size varies from 1.8-5.1 cm in length and about 0.4-1.5 cm in breadth near the mid body. The colour of body is generally pinkish. The margins are blackish or brownish due to the presence of vitelline glands. The alimentary canal is brownish in colour due to the presence of ingested bile of the host

Structure.

The *anterior end* of *F. hepatica* is triangular while *posterior end* is tapering. The anterior end terminates into a projection, which is called apical or cephalic cone or *head lobe*. At the apex of the head lobe is situated a small opening, the *mouth*, which is surrounded by an anterior muscular sucker, called oral or *anterior sucker*. At a short distance behind the head lobe is situated another saucer-shapted sucker known as *ventral sucker* or acetabulum, situated on the ventral side. The ventral sucker is superficial and has no connection with the interior of the body.

Fasciola hepatica was formerly known as *Distomum hepaticum* and the name of the group to which it belongs was known as *Distomidae*. These names were adopted as the mouth and anterior sucker were regarded as two suckers or two mouths (*di*, two ; *stoma*, mouth). The word *stoma* was used in trematodes for suckers, which are characteristic structures. But now it is certain that the mouth cannot be claimed as a sucker.

On the dorsal surface there is a single aperture in the middle line at about one-third of the length of the animal from the anterior end.

This aperture is the opening of *Laurer's canal.* This aperture appears during breeding season.

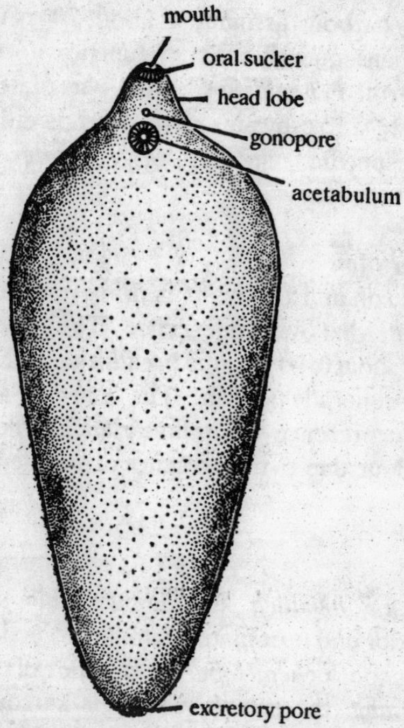

Fig. 3.1 *Fasciola hepatica.* External features.

Body wall

Body wall in Fasciola is characteristic and adapted to parasitic mode of life. It lacks a cellular epidermis and consist of the following layers :-

1. *Cuticle.* It is the outermost layer having stout, rough, blunt scales which project over the general surface. Cuticle is a secretion of *mesenchyme,* either of special type of cells called gland cells formed between muscle-fibres or of general mesenchyme. Cuticle is definitely non-chitinous and is made up of *scleroteins*. It is situated over the body and protects the liver-fiuke from the harmful effects of the different types of juices produced in the host. Below the cuticle is situated a thin transparent layer known as *basement membrane.*

2. *Musculature.* Immediately beneath the cuticle are three layers of muscles - the outermost circular layer, the medical longitudinal layer and the inner digonal layer. In addition to these muscles others make up the suckers and pharynx. Still others, the parenchymal muscles, are dorsoventrally oriented along the lateral regions of the body.

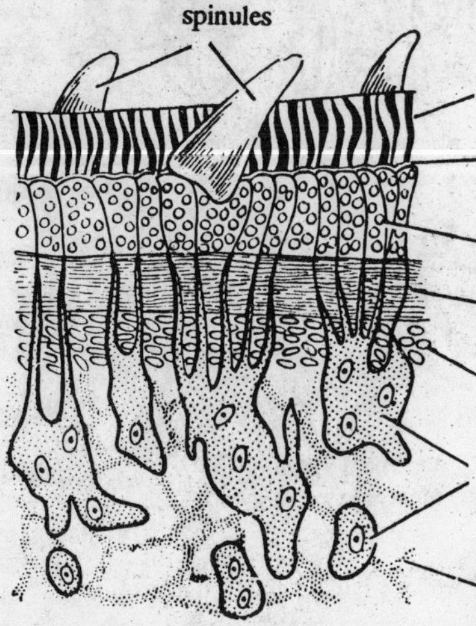

Fig. 3.2 *Fasciola.* V.L.S. body wall seen under light microscope.

3. *Parenchyna or mesenchyma.* Immediately beneath the musculature are numerous cells, known as *beta cells* or *mesenchymal cells*. Some of the beta cells possess cytoplasmic tubules that penetrate through the muscular layers and reach the cuticle. These specialize beta cells are thought to secrete cuticular materials that reinforce the existing cuticle.

Electron microscopic studies by *Threadgold* (1968) have revealed that the so called *cuticle* of Fasciola is actively a thick layer of multinucleated protoplasm containing mitochondria, canals of endoplasmic reticulum, vacuoles and vesicles. It is termed as *tegument.* It is continuous with *tegument secreting cells* lying embedded in the mesenchyme. The outer surface of tegument is thrown into numerous folds. These increase the surface area considerably for the absorption of fluids of the host. Further, tegument is perforated by fine *pore canals.* Through these canals fluids are absorbed into the mesenchyme.

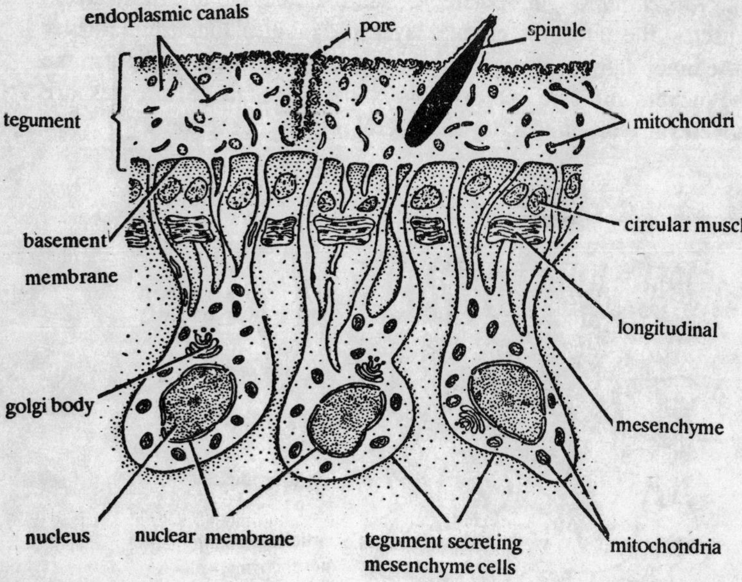

Fig. 3.3 *Fasciola.* V.L.S. body wall seen under electron microscope.

Functions.

The body wall of *Fasciola* is more than an inert protective layer, it is a dynamic structure of great importance in the metabolism of the organism. Although the animals derive a great deal of their nutrients through ingestion, certain chemicals such as certain amino acids, which are necessary as energy sources and as raw materials for the synthesis of complex body chemicals, are absorbed through the body wall. It is also possible that certain metabolic wastes are discharged through cuticle.

Digestive System

The digestive system is complex and highly branched with *mouth* but no *anus*. It consists of the following parts :

Mouth . It is a small opening situated ventrally at the apex of head lobe and surrounded by oral suker.

Oral chamber. Mouth opens into a short but funnel-shaped oral chamber.

Pharynx. It leads into a short, rounded, muscular pharynx with thick wall and pharyngeal glands.

Oesophagus . The pharynx communicates with a short-narrow oesophagus joining the intestine.

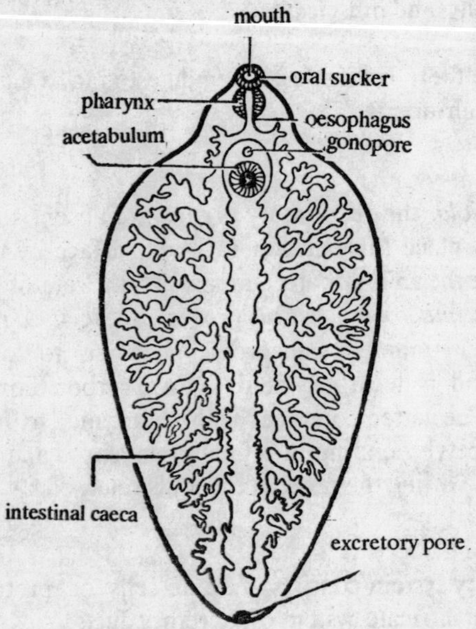

Fig. 3.4 *Fasciola.* Digestive system.

Intestine. The intestine immediately lbifurcates into two lateral branches, running one on either latera śide of the body upto the posterior end where these end blindly. A large number of caeca or diverticula are given out along either side of its whole length. The diverticula of the outer side are large and highly branched, while those of the inner side are short and unbranched.

Food and feeding. The food of Fasciola is bile and blood as it lives in the bile passage of sheep and cattles. It takes about 0.18-0.2 ml of blood per day. Small microscopic debris is also taken as food. The food is taken by the sucking effect of the pharynx and is pumped into the intestine.

Digestion. The exect manner of food digestion is not well understood. Digestion is supposed to be extracellular, carried in the intestinal ceaca. Proteolytic enzymes are reported in the animal. The digested food is diffuses into the surrounding mesenchyme. The mesenchyme distributes digested food to every part of body. As earlier mentioned the tegument also helps in the absorption of certain amino acids as well as monosaccharides (glucose and fructose). The reserve food material is in the form of glycogen and fat which are stored in mesenchymal cells and muscles.

Egestion. As there is no anus, the undigested food materials are egested out through mouth.

Respiration

In the *Fasciola,* the respiratory organs are absent so that the respiration takes place through general body surface by the process of diffusion. Since the animal is an endoparasite, the rate of respiration is very low and is *anaerobic*. In this process, there is no necessity of oxygen. Glycogen undergoes anaerobic glycolysis to form pyruvic acid. Pyruvic acid is decarboxylated to form carbon dioxide and an acetyl group. The latter combines with coenzymes to form acetyl coenzymes A. Acetyl coenzyme A is finally condensed and reduced to form fatty acids. During the process energy is released.

Excretory system

The excretory system consists of flame cells or protonephridia connected with an intricate system of excretory ducts.

1. *Excretory ducts*. A single and median longitudinal excretory canal which runs just behind the ventral sucker and terminates into the *excretory pore*. It is ramified throughout the body. In the anterior region it receives two ventral and two dorso-lateral excretory canals. The dorsal ducts before opening into the main longitudinal excretory canals are united so as to form a common duct. Main excretory canal as well as dorsal and ventral-lateral excretory canals divide into small capillaries or tubules, each of which ends internally into a *flame cell.*

2. *Flame cells.* The flame cells are characteristic, each has a thin elastic wall with a nucleus and a cavity containing many long cilia arising from basal granules or blepharoplasts. The cilia in a living animal vibrate like the flickering of the flame, hence named as flame

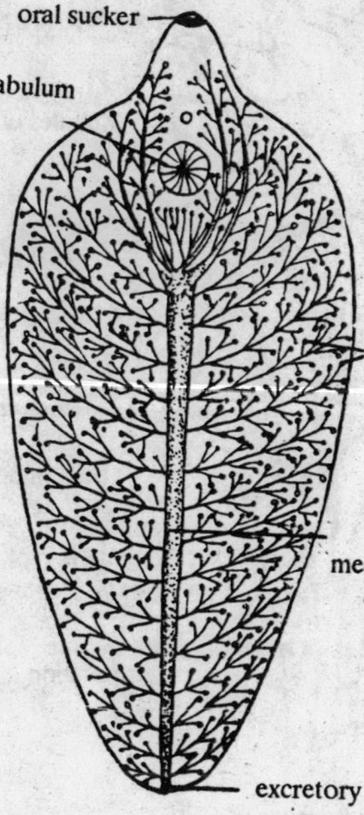

Fig. 3.5 *Fasciola*. Excretory system.

cells. The cavity of flame cell is continued with the cavity of capillary. The shape of these cells is irregular due to preudopodial processes into the mesenchyma.

Physiology of excretion

The excretory products are carbon dioxide fatty acids and ammonia. These are diffused into the flame cells from the surrounding mesenchyme. The fluid moves through the tubule by the action of cilia into the excretory ducts and finally passes out of the body through the excretory pore. All the ducts except the median longitudinal duct are lined with cilia. The course of the flow of excretory products is as follows :

Excretory products --> mesenchyme --> flame cells
↓
main longitudinal canal <-- vessels <-- twigs <-- Capillaries
↓
Excretory pore --> outside.

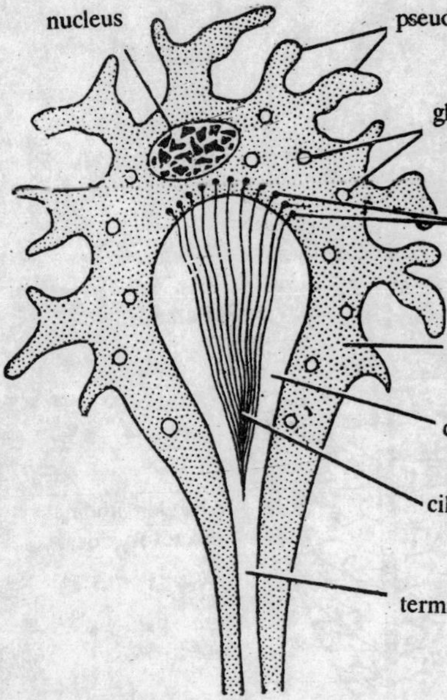

FIg. 3.6 *Fasciola*. A flame cell.

Nervous System

The nervous system is well developed in spite of the fact that due to sluggish and parasitic habitats such a large correlation is not need. The nervous system includes :

(a) *Central nervous system*. Two cerebral ganglia are prominant masses of nervous material joined together by a nerve ring around the oesophagus. From each cerebral ganglion arise three longitudinal nerve cords, dorsal, ventral and lateral. Of these the lateral nerves are much longer than the other. They alone reach the posterior end of body. They are interconnected at intervals by transverse commissures.

(b) *Peripheral nervous system*. From each cerebral ganglion three nerves arise anteriorly and innervate the head lobe and the oral suckers. Fine nvërves arise from nerve cords all along their course and supply the various parts of the body.

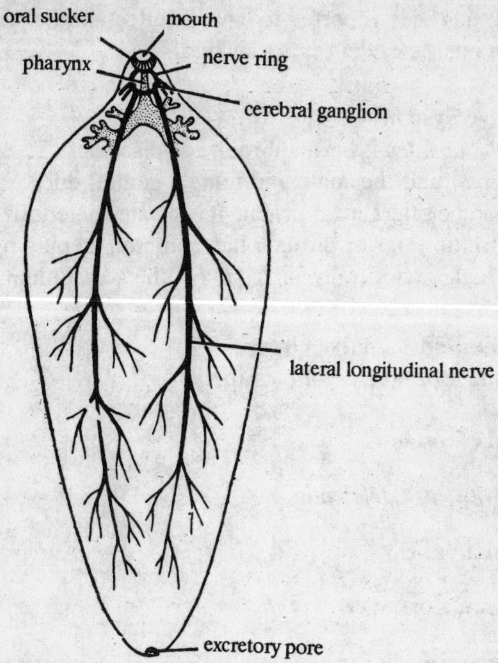

Fig. 3.7 *Fasciola.* Nervous system.

Sense organs

Due to parasitic mode of life, special sense organs are absent. It, however, possesses tectile organelles or *Tangoreceptors*. These are bulb-like nerve endings. These are abundant in the sucker region.

Hormones in Fasciola

In *Fasciola* serotonin is present and it seems to play a role in the tissues comparable to that played by adrenalin in mammalian tissue. It is not known whether serotonin acts as a humoral component of the nervous system, as it is reported to do in some other animals. The tissues of *Fasciola* contain an acetylcholine-like substance and cholinesterase ; the specificity of cholinesterase for acetylcholine has not been established, but it has been suggested that these components are related to activity of nervous system.

Insulin has been reported to have an effect on the carbohydrate metabolism but the evidences are lacking.

Reproductive system

Fasciola hepatica is hermaphrodite or disocious. The gonads are well developed and the male and female genital ducts open into a common chamber, the genital atrium. It is situated anteriorly in the body and opens to the exterior through the common genital aperture or gonopore, located ventrally in front of the acetabulum.

1. The Male Reproductive Organs

The male reproductive organs are :

(i) A pair of testes

(ii) A pair of *vasa deferentia*

(iii) Seminal vesicle

(iv) An ejaculatory duct

(v) Cirrus and cirrus sac.

(i) *Testes*. These are greatly ramified and tubular structures situated in the middle part of the body, one behind the other (*tandem arrangement*).

(ii) *Vasa deferentia*. A narrow delicate duct, the *vas deferens,* emerges out from each testis and runs forward to meet the fellow of opposite side forming the common sperm duct just below the ventral sucker.

(iii) *Seminal vesicle*. The common sperm duct immediately leads into a large muscular pear-shaped sac, the *seminal vesicle*, which lies in front of ventral sucker.

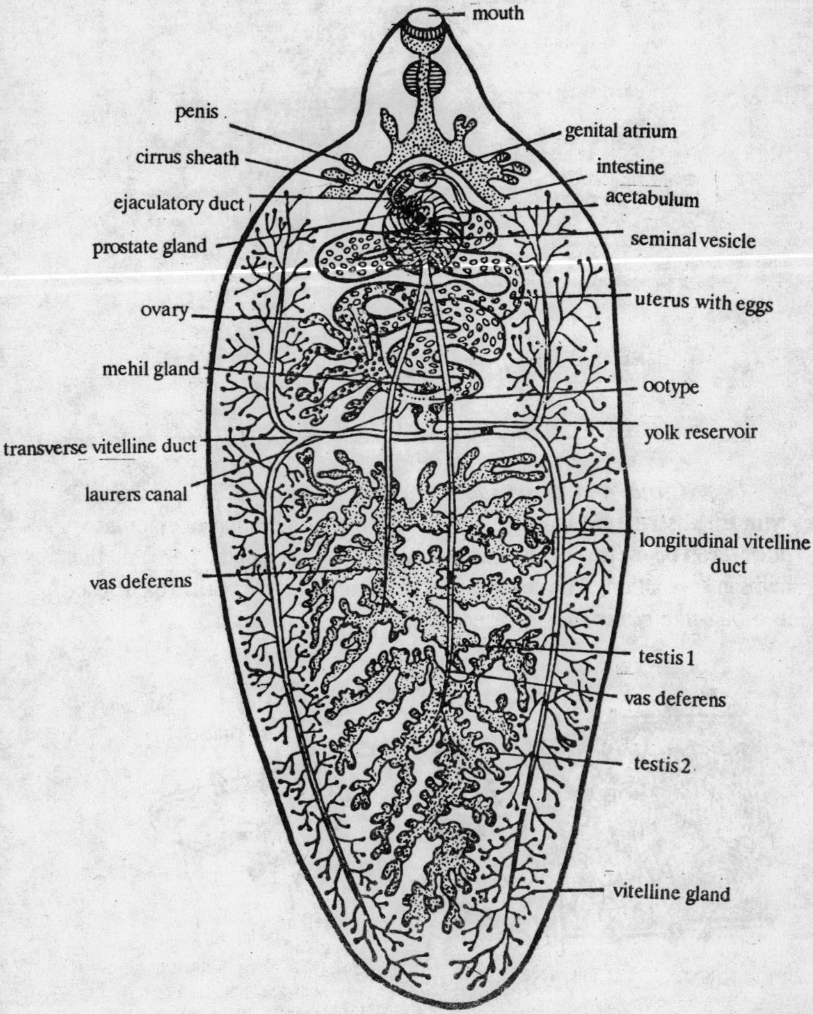

Fig. 3.8 *Fasciola*. Reproductive system.

(iv) *Ejaculatory duct*. The ejaculatory duct arises as a fine convoluted tube from the seminal vesicle and runs forward in a zigzag fashion through the cirrus to open into the *genital chamber* through the *male genital pore*.

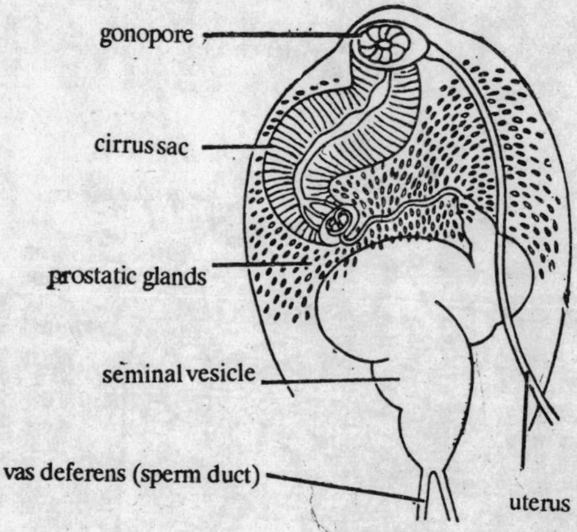

Fig. 3.9. *Fasciola.* Details of male reproductive system.

(v) *Cirrus and cirrus sac.* The cirrus or penis is a muscular cylindrical structure traversed internally by the passage of ejaculatory duct. It can be everted out and drawn in through the genital pore and thus helps in copulation. The cirrus and the seminal vesicle both are enclosed in a bag-like *cirrus sac.*

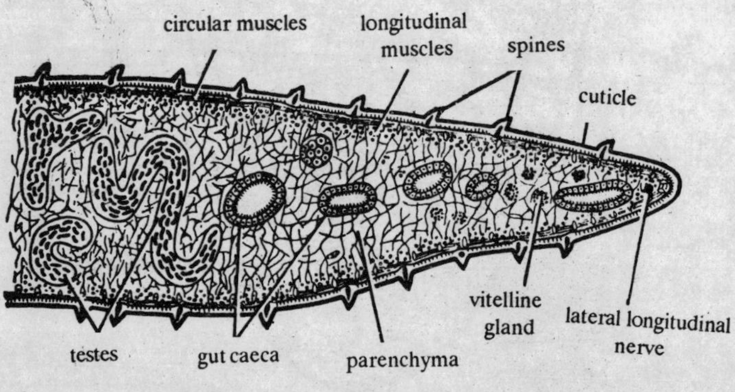

Fig. 3.10. *Fasciola.* T.S. body through male reproductive system.

(vi) *Prostate glands.* Numerous unicellular prostate glands are found around the ejaculatory duct.

(B) Female Reproductive Organs

The female reproductive organs are :

(i) Ovary
(ii) An oviduct
(iii) Uterus
(iv) Laurer's Canal
(v) Mehli's gland
(vi) Vitelline glands and vitelline ducts

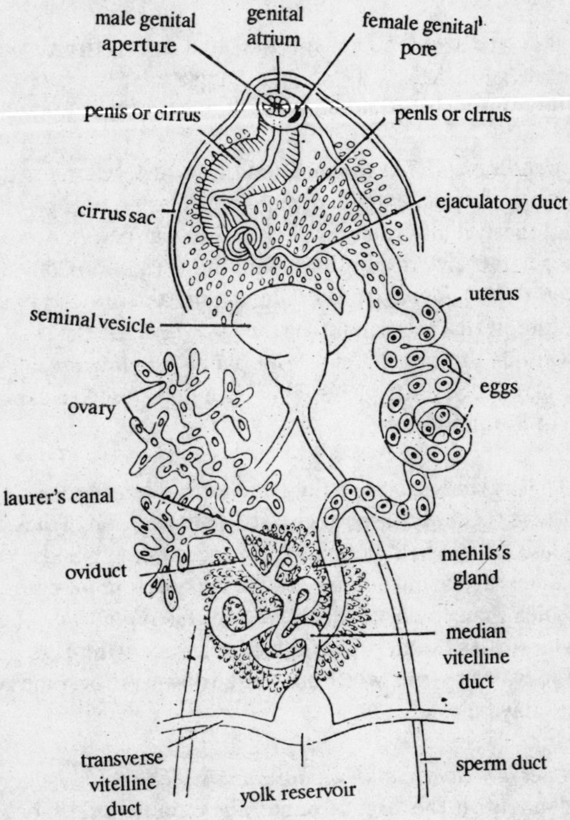

Fig. 3.11. *Fasciola.* Details of male and female reproductive system.

(i) *Ovary.* The ovary is a large, highly branched tubular structure which is situated on the right side in front of testes in the mesenchyme. It occupies the anterior one-third of the body.

(ii) *Oviduct*. From the inner side of ovary arises a short, narrow and convoluted duct, the *oviduct*. It runs down ward to join the uterus.

(iii) *Uterus*. It is a long wide and highly convoluted tube that extends upto the genital atrium, opening into it through the female genital aperture, close to male genital pore on the left-side. It contains a large number of capsules containing fertilized eggs.

(iv) *Laurer's canal*. The laurer's canal arises from the oviduct and acts as a sperm duct. It develops a temporary opening on the dorsal body surface during breeding season. It may serves as vagina.

(v) *Mehli's gland.* At the junction of oviduct, vitelline duct and uterus a mass of unicellular gland lies known as *Mehli's glands*. The function of these glands remains unclear. It was believed that these glands were merely responsible for the formation of shell but *Stephenson* (1947), *Rennison* (1953) and *Johri* and *Smyth* (1955) have shown definitely that this cannot be the case, because the shall of eggs is rich in various proteins, phenols and phenolases that are not present in these glands. Several views have been proposed to explain the function of Mehli's glands:

(a) These glands secrete a fluid that enchances the hardening or tanning process of newly formed eggs. (b) The secretion of these glands cause release of the shell globules from the vitelline glands. (c) The secretion forms a thin membrane around the cells forming eggs, and the shall globules than built up from within this membrane (d) The secretion lubricates the uterus facilitating passage of the eggs. (e) The secretion activates sperms which are passed down. All or some of these hypothesis may be valid.

(vi) *Vitelline glands and vitelline ducts*. The vitelline glands or the yolk glands are in the form of numerous minute vesicles or follicles arranged on either side of the body from one end to the other. A fine duct arises from each vesicle and after uniting with similar ducts from other vesicles opens into the *lateral longitudinal ducts*. The two longitudinal ducts are connected in the middle line by a *transverse vitelline duct*. A medium duct arises mesially from the transverse vitelline duct and joins the oviduct.

The functions of vitelline glands is now know. In addition to contributing yolk material for incorporation within the egg, the vitelline glands also secrete large globules known as *shell globules* - which envelop the developing egg and eventually coalesce and become hardened to form the shall. This hardening process involves the tanning of protein present within the coalesced globules by quinone. Thus vitelline glands contribute both yolk and shell material.

Phenolase

Phenol -----------------> quinone ---------> Tanned protein

Protein -----------------> protein ↗

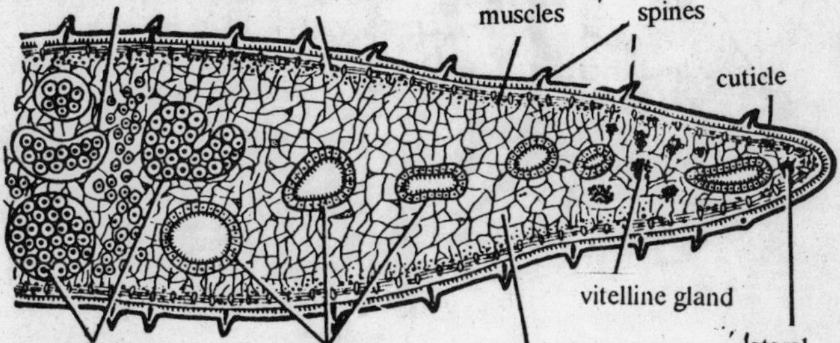

Fig. 3.12. *Fasciola*. T.S. body through ovary and uterus.

Ootype. In some cases of the oviduct after receiving Laurer's canal and a common vitelline duct swells to form ootype. In *F. hepatica* according to *Stephenson* (1947) the ootype is absent.

LIFE HISTORY

Copulation and fertilization. Cross - fertilization takes place by insertion of cirrus of one fluke into the Laurer's canal of another fluke. Transference of biflagellate sperms directly into the oviduct is possible. Secretion of prostate glands and as mentioned above, the secretion of Mehli's gland keep the sperms active for fertilization. Laurer's canal become invisible after copulation. Fertilization takes place in oviduct. Self-fertilization is also seen here the sperms enter the uterus of same fluke through female genital pore and pass down to reach oviduct where fertilization takes place.

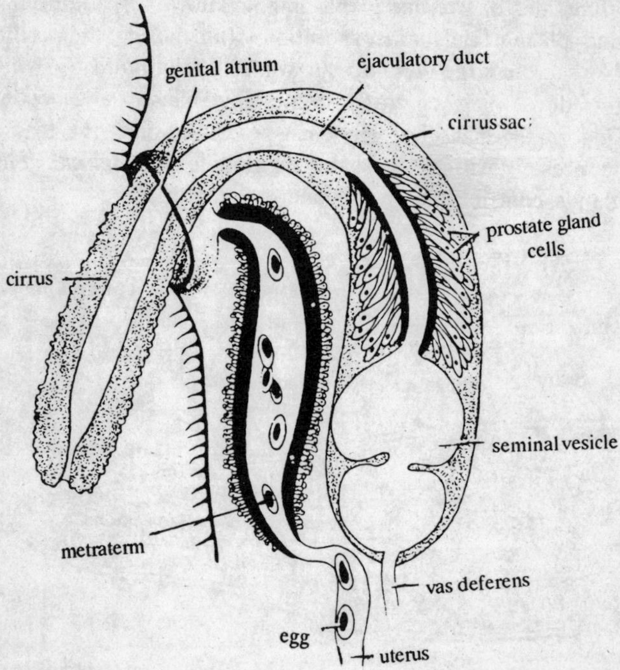

Fig. 3.13. *Fasciola.* Shows cirrus coming out of gonopore.

Fertilized egg gets about 30 or so yolk - cells and shell globules from vitelline glands by the way of various vitelline ducts. The shell is formed with a lid or *operculum* on one side. *Rowen* (1956, 57) demonstrated that there is a viscous granular cushion at the opercular end of the egg, situated immediately beneath the operculum. The zygote and a mass of yolk - cells enveloped by the shell constitute the capsule. When complete the *capsule* passes into the uterus.

Capsule formation occurs throughout the year and one fluke may produce about 50,000 capsules. A single fluke may produce a total of 2,00,000 capsules in about 11 years.

The capsules differ in shape and size in different animals. Hence, with the size and shape of capsules the parent flukes are recognized.

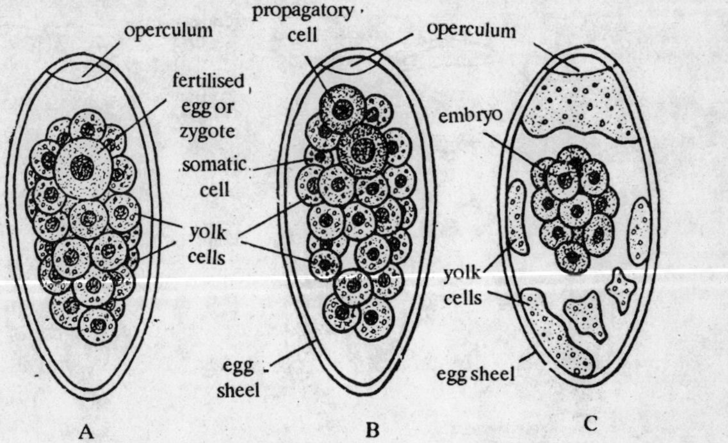

Fig. 3.14. *Fasciola*. Early stages of development. A = Fertilized egg, B = Two cell-stage, C = Many cell stage.

Development. Clevage starts immediately after fertilization. The fertilized egg divides into a small granular *propagative cell* and a large ectodermal *somatic cell*. Somatic cell divides and gives rise to the body-wall of the larva. Propagative cell later on divides into a cell similar to *somatic cell*, which takes part in the development of mesoderm, endoderm, and a cell called *germ cell* afterwards reaches the posterior part of the body and forms germ-ball consisting of 14 to 20 germ cells.

Capsules then escape through the gonopore and are liberated into the bile capillaries from where they reach the intestine of the host and are passed out with the faeces.

The development remains arrested while the embryos remain in the faces, they may survive in wet faecal matter for several months. If washed free, the development of the embryo proceeds. The optimum temperature for development ranges from 10 - 30°C. At 30°C the encapsulated embryo differentiates into *miracidium larva* within eight days. The availability of oxygen is also a factor. The fully formed miracidium is a more or less conical little animal covered with ciliated epithelium. It does not look at all alive an adult fluke but does show resembelance to some free-living flat worm.

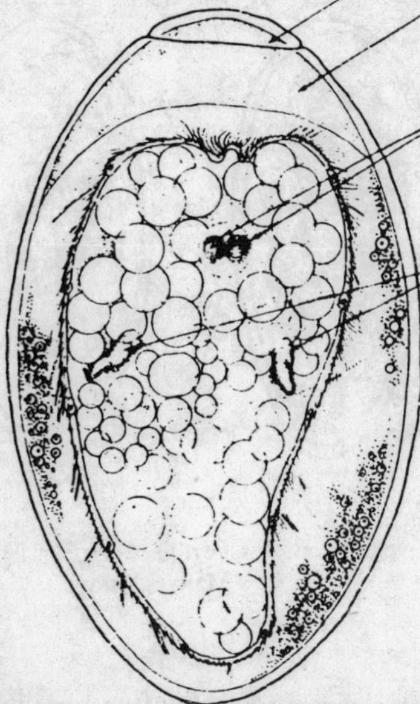

Fig. 3.15. *Fasciola.* Miracidium just before hatching.

Miracidium larva. As the miracidium is formed the hatching enzyme is secreted in response to exposure to light. The enzyme dissolves the cementing material by which the operculum is attached, thus releasing the operculum. Expension of the granular cushion, accompanied by exosmosis of salts and other materials from within the egg pushes off the operculum. The hatching occurs in water and free-swimming miracidium penetrates the intermediate host.

Structure of miracidium. It is an elongated, about 0.07 mm long creature with a broad anterior end and a narrow posterior end. Anteriorly there is a *conical* or *apical papilla* or terebratorium. It is cliated except in the region of apical papilla. However, anterior cilia are slightly longer.

The ectoderm is formed by 21 hexagonal plate like cells which are arranged in 5 rows. The I row consists of 6 small plates. The II row contains again 6 plates, III row has 3 plate cells, 4 plates in IV row and 2 plates in V row. The arrangement of the ciliated plates is quite constant and is of considerable value in determining phylogenetic relationships. In *F. hepatica* the cells of first row arranged are two dorsal, two ventral and two lateral in position. The cells of second row are three dorsal and three ventral. In the third row the cells are one dorsal and two ventero-lateral. The fourth row has two right and two left cells and the fifth row has one left and one right cells.

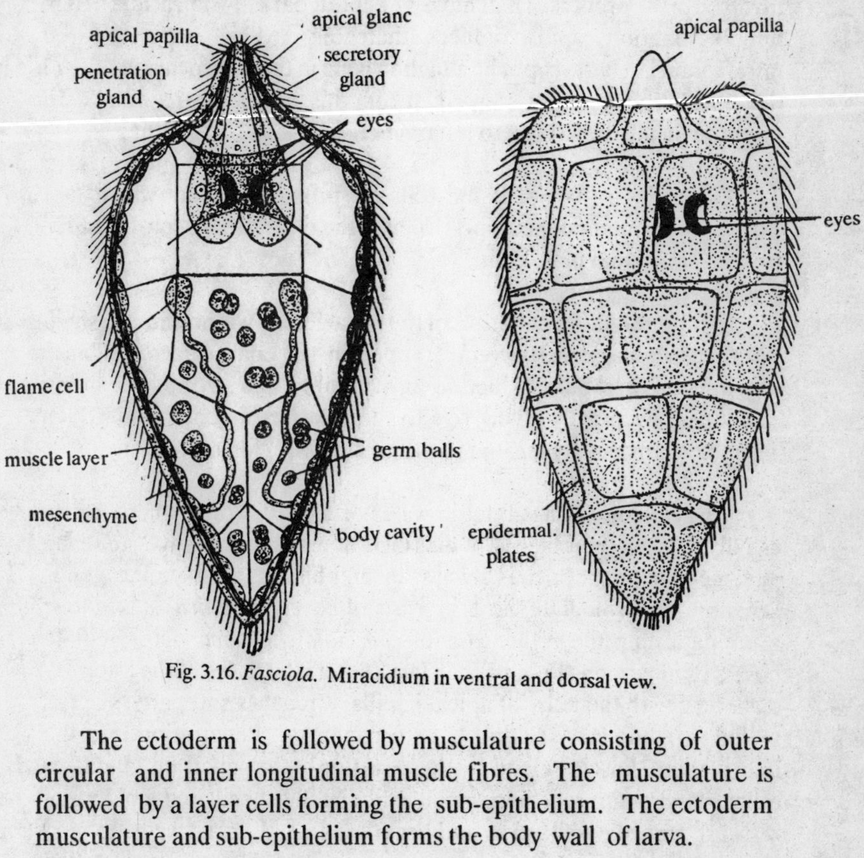

Fig. 3.16. *Fasciola*. Miracidium in ventral and dorsal view.

The ectoderm is followed by musculature consisting of outer circular and inner longitudinal muscle fibres. The musculature is followed by a layer cells forming the sub-epithelium. The ectoderm musculature and sub-epithelium forms the body wall of larva.

At the anterior tip of the miracidium is found a mobile apical papilla. A pair of cephalic *glands* or *penetration glands*, located in the mesenchyme, empty to the outside near the apical papilla. The secretion of these glands aids in dissolving the host's tissue during penetration process. Besides a sac-like multinucleated *apical gland* is located between the two cephelic glands. Previously it was supposed to be the gut of larva.

The sensory organs of miracidium includes two eyespots. Eyes contain pigments and are situated at the surface of the second row of epidermal plate over the brain. The eyes may be primary receptors involved in the process of hatching. I7 kept in dark few miracidia hatch, but illumination rapidly induces hatching. The *brain* of the miracidium is a large cephalic ganglion lying in the parenchyma behind the apical region. It is X-shaped. From this nerve centre, fibres are oriented in all directions to innervate the various body tissue.

There is a pair of flame-cells or protonephridia, which are excretory organs and their ducts open to the outside between IV and V rows of epidermal plate cells.

During the differentiation of the miracidium within the egg shall, certain germinal cells become trapped in the parenchyma. These develop into germ balls by increasing in number and size and becoming enveloped, eventually giving rise to the next larval generation. Such germ balls are usually located in the posterior region of the body.

Miracidium swims about in water or moisture film, where cattle usually pass out their faeces. It dies in 30 hours unless it is able to reach intermediate host, which is always an amphibious snail of the genus *Limnaea*. Intermediate host in case of *Fasciola indica* is *Limnaea* acuminatisata while in the case of *Fasciola hepatica* intermediate host is Limnea truncatula. When larva enters the tissue of the intermediate host with the help of apical papilla, it reaches soft parts, such as mantle, pulmonary sac and hepato-pancreas. Penetration into the intermediate host is helped by the secretion of the penetrant glands or other glands. On reaching the intermediate host, ciliated epidermal cells are cast off, while eyes and brain are lost. Larva swells up and changes in shape and size, so as to become a sac-like structure, called sporocyst. Germ-balls of miracidium do not disintegrate.

Sporocyst. The sporocyst is an elongated, hollow, sac-like larva. Its body wall consists of non-cellular cuticle followed by musculature and sub-epithelium. The brood chamber is hollow space in the centre of the sporocyst within which are found the germ balls that eventually give rise to the next generation. Sporocyst contains no alimentary canal, nervous system or reproductive systems. The protonephridia are present opening laterally by pores.

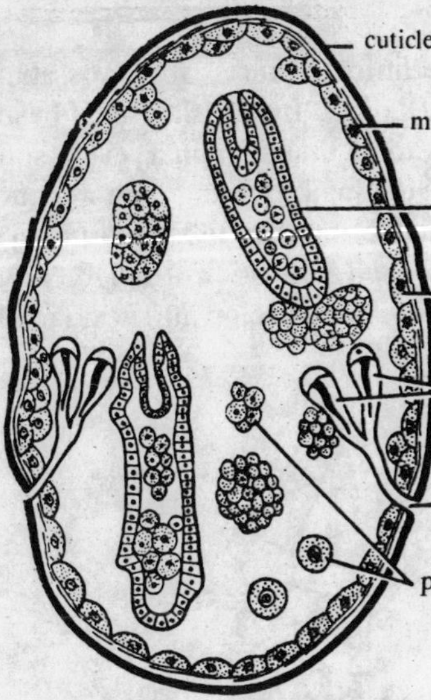

. Fig. 3.17. *Fasciola.* Sporocyst

The sporocyst moves about by muscular contractions in the host tissues, absorbing nourishment from them and causing them a considerable damage. Its germ balls develop into the next larval forms, the redial. Each sporocyst produces 5-8 redial. Some of the germinal cells are again set aside within the redial for the formation of the next larval forms.

Redia larva. The rediae emerge from the sporocyst by the rupture of body ball of sporocyst. Each radia is an elongated and cylindrical form, measuring 1.3-1.6 mm in length. In has a circular ridge, called collar near the anterior end and a pair of projections termed *lappets* near the posterior end. The body-wall consists of the usual cuticle, musculature and subepithelium. Unlike the sporocyst, the redia possesses a mouth located at the anterior terminal of the elongated body. The mouth leads into a muscular pharynx, which in term leads a blind-sac, the gut. Numerous unicellular pharyngeal glands open into the pharynx.

Protonephridia divide further and form a much branched system. All the flame cells of one side open out through a common excretory duct. A birth pore is present, lateral to and in proximity of the mouth. Within the brood chamber are found germ balls.

Redia is a very active little creature. It moves about by the contractions of the body and lappets. It nourishes itself by sucking fluid and cells of the host tissues with its muscular pharynx. It soon migrates into the digestive glands of the snail. During summer months when the food is sufficiently available, mother redial mature in 12-18 days, the germ balls give rise to a second generation of daughter redial. Germ balls of daughter redial in the winter develop into next larval stage, the cercariae.

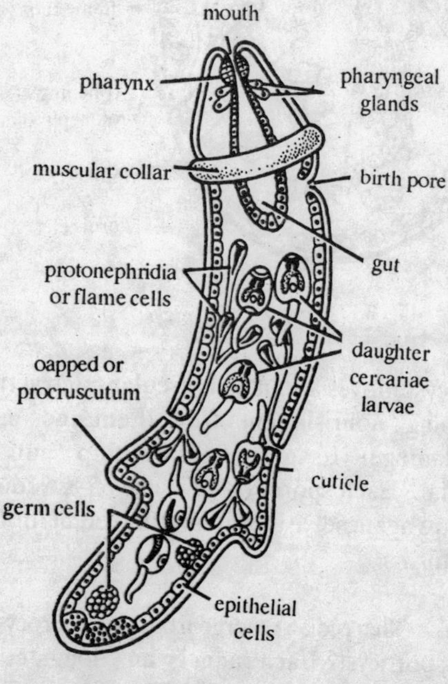

Fig. 3.18. *Fasciola.* Redia.

Cercaria larva. Cercariae leave the mother or redial through their birth pores. From each redia 14-20 cercaria are produced. Cercaria is about 0.25-0.35 mm long and is an oval larva with a tail. The body wall consists of cuticle, musculature and subepithelium. On the anterior side, there is a mouth with a sucker, a sectorial pharynx, long oesophagus and a bifid intestine. A ventral sucker or acetabulum is present on the ventral side. A large number of dark, brown cells, called *cystogenous glands* are found beneath the cuticle. There is a pair of excretory ducts, each with several flame cells. These open near the hind end of the larva into a bladder that leads out through a pair of nephridiopores. A nervous system similar to that of adult has also appeared. Rudiments of reproductive organs develop from the germ cells near the hind-end.

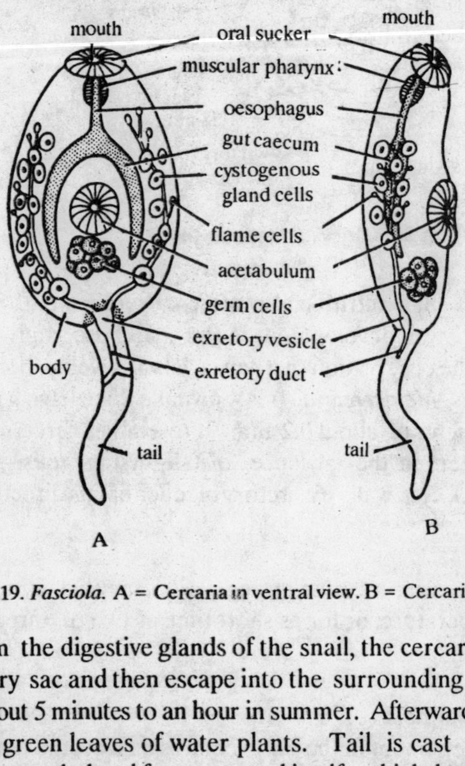

Fig. 3.19. *Fasciola.* A = Cercaria in ventral view. B = Cercaria in lateral view.

From the digestive glands of the snail, the cercariae pass into the pulmonary sac and then escape into the surrounding water. Cercaria swims about 5 minutes to an hour in summer. Afterwards, it settles down on some green leaves of water plants. Tail is cast off and the body becomes rounded and forms around itself a thick brownish cyst wall from the cystogenous glands. The encysted cercaria is now called metacercaria.

Factors affecting development in Snail

The rate of development in the snail host is a function of the environmental temperature. *Kendell* reported a curious effect of temperature on the development of *Fasciola.* In snails maintained at laboratory temperature, daughter redial were not produced, the first generation of redial giving rise to cercariae. On the other hand snails exposed daily to a period of low temperature i.e. 4-5°C, daughter redial were produced from redial.

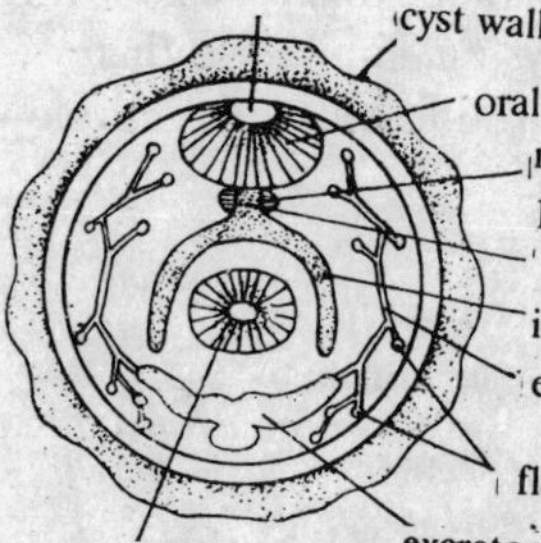

Fig. 3.20. *Fasciola.* Metacercaria.

The state of snail nutrition also affects development. The effect of food availability on development in the snail host is probably the explanation for the observation that fewer larval develop in small snails than in large ones, *Metacercaria*. It is a juvenile fluke having a rounded form with a diameter of about 0.2 mm. It resembles cercaria larva but differs from later in the absence of tail, cystogenous glands and presence of thick cyst wall. Excretory bladder opens directly through a single pore.

The metacercaria, encysted on vegetation, may live for as long as a year at low temperature, or for as short time as two or three weeks at 25°C.

Inspection to final host

The final excystment in the host requires an elevated temperature i.e. 38°C, low redox potential, a high carbon dioxide tension, and the presence of bile. The factors necessary for the excystment of *Fasciola* matacercariae seem to serve as stimuli to activate the worms, but the mechanism is not clearly understood. When metacercaria reaches the intestine of host, the cyst is digested and young fluke is liberated. The young worm penetrate the wall of gut and passess into the peritoneal cavity. In the body cavity the young worms seem to move about rather aimlessly until the encounter the capsule of liver. On contact with the liver, the flukes burrow into it, devouring liver parenchyma as they move. It grows in size 5-6 weeks and finally settles in the bile duct. It attains sexual maturity and begins to lay eggs 11-13 weeks after ingestion by the mammalian host.

Nature of life-history

The life-history of *Fasciola hepatica* is much complicated due to the occurrence of several larval forms in a series. Formarly, it was held that the redial and cercariae develop by parthenogenesis from the propagatory or germ cells, which were believed to be the eggs (*Grobben*, 1882). This kind of asexual parthenogenetic reproduction by larval forms is known as *heterogamy*. This view was widely accepted for sometime and under its sway, three terms were coined. These are *parthenitae* for sporocyst and rediae larvae, *adolescariae* for cercaria larvae, and maritae for adult flukes. This view, however, is now considered erroneous and has been given.

It has been found that the germ cells represent in the larval forms are not eggs. Instead they are diploid cells derived by the mitotic divisions of propagatory cell which is separated from the zygote at its first cleavage. Thus it can be said that they are parts of zygote. It, thus, follows that one zygote produces several larval. Thus, this process of reproduction in sporocysts and rediae has been looked upon as *polyembryony* by *Ishii* (1934), *Chen* (1937), *Rees* (1940) and *Cart* (1944). Since all the larval resulting from a zygote are not formed simultaneously but at intervals, the phenomenon should be called delayed polyembryony.

Steenstrup (1942) and some other workers interpret polyembryony as a sexual multiplecation and hold that there is a alternation of sexual and asexual generation in life-history of *Fasciola*. But *Hyman* is of the view that it is a continuous ontogeny involving asexual multiplication in the larval stages.

Pathogenesis

Pathogenesis as a direct result of the flukes' activities may be either chronic or acute. Acute *fascioliasis* or *liver rot* occurs during the preadult migration of the flukes in the parenchyma of the liver, and sometimes other organs, for about 8 weeks. The most striking changes in acute fasciolosis are seen in the liver, where there are haemorrhagic flukes. The liver becomes Swollen and in some instances, may become so swollen that the capsule ruptures. The by-products produced by worms are very toxic and cause *anaemia, diarrhoea, eosinophilia* etc.

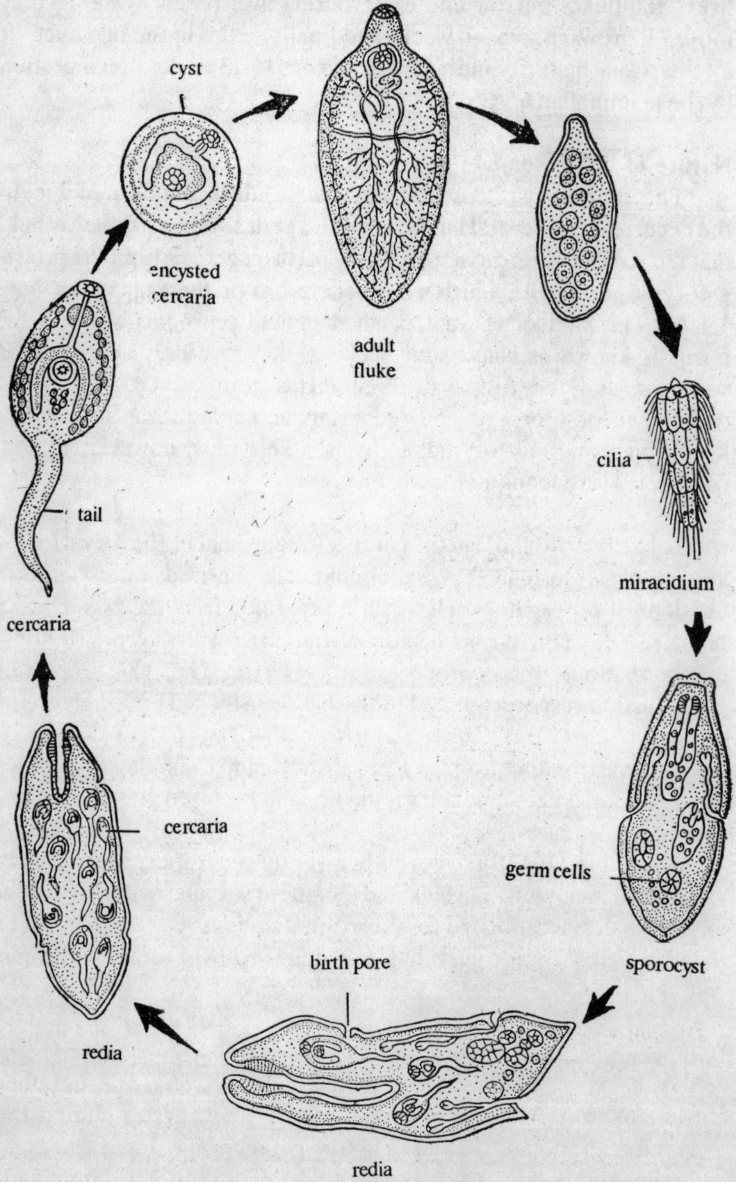

Fig. 3.21. *Fasciola.* Life - cycle.

Chronic fascioliasis occurs beyond 12 weeks, when the flukes have reached the bile ducts and are maturing sexually. Although milk and wool production looses are the usual results of chronic infections, deaths also occur.

The liver performs a vast array of essential functions in vertebrates, and it is axiomatic that liver damage will have a variety of detrimental effects. The adult fluke causes anemia, haemorrhage and supression in production of new red blood corpuscles. The result of liver damage is failure to gain weight in young animals, weight losses in older animals, reduced milk production and lowered fleece weights in sheep. The osmotic pressure of the blood is lowered and fluid is retained in the tissue spaces.

Treatment. Carbon tetrachloride has been used in sheep at 500 mg/kg. Hexachlorethane was developed especially for use in cattles. Other drugs used are emetine hydrochloride, filcin, tetrachloroethane, etc.

Control. Control of liver flukes is effected in three major ways :

1. Snail control.
2. Treatment of infected sheeps.
3. Herd management through predicting out breaks of disease.

Parasitic adaptations in *Fasciola*

The main changes that occur in the structure and life-cycle of a parasite, with the help of which it is capable of leading a parasitic life successfully, are called the *parasitic adaptations.* Since *Fasciola* is an endoparasite in the liver of sheep, it has undergone similar modifications under the stress of the parasitic mode of life which can be studied under two heads:

I. Structural modifications.
II. Modifications in the life-history.

I. Structural modifications

These modifications are of two types.

(A) Morphological.
(B) Anatomical

(A) Morphological modifications

1. Body is dorso ventrally flatenned leaf-like and is nearly triangular due to which the danger of injury to its body is minimized.
2. Adhesive organs or suckers are present, an *oral* and a ventral or *acetabulum*. These are muscular and help in attaching to the tissue of host.
3. External features simple and is not divisible into head, trunk and tail. Appendages are also absent.

(B) Anatomical modifications

1. Body is devoid of cellular epidermis, thick cuticle is present to prevent the animal from the effect of chemicals from host body. Cilia absent in adult. Cuticle is provided with spinules for attachment.
2. The thickening of cuticle in the oral and ventral regions leads to the formation of two suckers.
3. The locomotory organs are lacking except in free living larval stages as adult *Fasciola* is and endoparasitic and it need not make excursions for food and safety.
4. The digestive system is incomplete. The *mouth* is anterior and terminal and surrounded by oral sucker. The pharynx is also muscular and sectorial. Bother are adapted to suck the nutrient.
5. Intestine is highly branched so that digested food can be distributed to different parts of body. Anus is absent.
6. Circulatory and respiratory systems are absent. Anaerobic respiration takes place.
7. Excretory system is well developed. It consists of extremely branched excretory canals which terminate into flame cells.
8. Nervous system is poorly developed. Sense organs are absent.
9. Reproductive organs are extensively developed. The highly branched ovary and tests are capable of producing large number of gametes. Bisexuality offers full chances of fertilization.

II. Modifications in life-history

1. The production of eggs in a very large number, about 10 million, due to which the threat of survival of parasite is removed.
2. Eggs are shelled and provided with an operculum.
3. A secondary host is present for the dispersal of species.
4. Multiplecation takes place by *paedogenesis* during larval stage.

5. Miracidia are free swimming larval stages because they have to search new hosts. The cercariae are encysted when out of the body of secondary host which ensures protection from external effects.

4

SCHISTOSOMA

The schistosomes are members of order Digenea, but are different from most members of order in that the sexes are separate. The term *Schistosoma* means split body and refers to the fact that the males have a ventral groove called *gynecophoral canal* in which females lodge. Adults of the family Schistosomatidae are found in the blood vessels of vertebrates, both birds and mammals. Three species namely *Schistosoma mansoni, S. japonicum* and *S. haematobium* are found in the blood of man and cause *Bilharziasis* or *Schistosomiasis.* No other disease caused by animal parasite causes as much harm to man as *Schistosoma* does with the sole exception of *plasmodium* causing malaria. Schistosomiasis was discovered in 1779 in France. In 1851, *Bilharz* reported *S. haematobium. Harley* and *Cobbald* (1864) reported that some molluses are intermediate host. The molluscen hosts are species of *Tropicorbis, Biomphalaria* and *Australorbis.*

SYSTEMATIC POSITION

Phylum	-	**Platyhelminthes**
Class	-	**Trematoda**
Order	-	**Digenea**
Family	-	**Schistosomatidae**
Genus	-	***Schistosoma***

Distribution

The human schistosomes are distributed through some of the countries of the world at the lower end of the economic scale, or at least the parasites are found in relatively poor rural people. *S haematobium is* found in Africa especially in Egypt and Sudan where up to 90% of the rural masses may carry the infection. The intermediate host is the freshwater snail *Bulinus truncatus* and related species of *Bulinus. S. japonicum* is found in the Far East (China, Japan, Philippines, Thailand,

Vietnam etc.). The mollusc host for *S. japonicum* are species of the snail genus *Oncomelania.*

S. mansoni is found in many parts of Africa (including Egypt) and Tropical America including West Indies.

Habits and habitat

They live in the blood vessels of the bladder, abdominal cavity etc. *S. mansoni* found in the portal blood of the host, usually in the smaller branches of the inferior mesenteric vein in the region of the lower colon. *S japonicum* adults inhibit the branches of super mesenteric vein in the proximity of the small intestine. *S. haematobium* adults reach the vesical, prostatic and uterine plexuses by way of inferior haemorrhoidal veins.

External morphology

The adult worms are beautifully adapted to life in blood vessels. The mature worms, male and female, live in *copula* (in almost permanent copulation).

Colouration. The colour of the body is greyish or pinkish. The alimentary canal presents reddish colour due to degenerating blood cells of host.

Shape and Size. The body is thin, cylindrically elongated. The male are comparatively thicker and small than females. In *S. mansoni* the male, measuring 6.4-9.9 mm in length, is broader then the female. The female measures 7.2-14 mm in length. In *S. japonicum* the males may reach upto 22 mm in length while females 26 mm in length. In S. *haematobium* the adult males may become 15 mm long and the females 20 mm in length.

Suckers. Both male and female possessing two suckers each. Both the suckers are present and are located in the anterior portion of the body. The oral sucker surrounding the mouth and the acetabulum present on ventral side lies close to the oral sucker. Acetabulum is very powerful in the males.

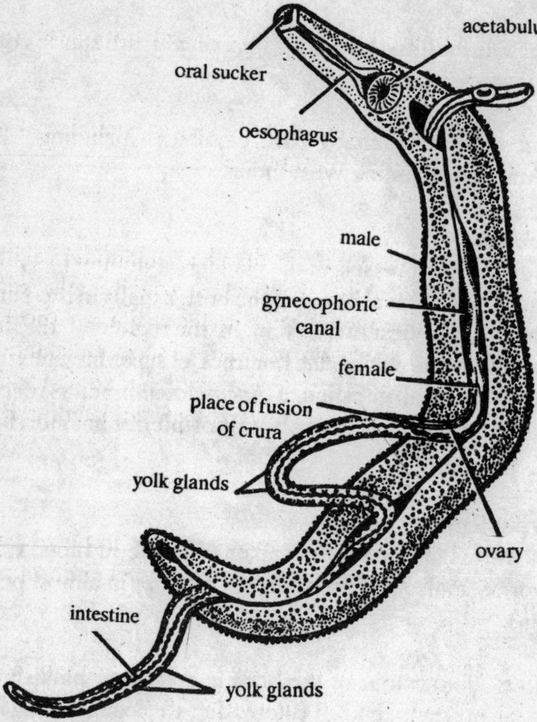

Fig. 4.1. *Schistoma.* Male and female.

Tubercles. In S. mansoni minute sensory tubercles cover the body surface of the male, and microscopic tufts of cilia are found on these protuberances. In *S. haematobium* the tubercles are very fine. In *S. japonicum* these are absent. *Gyanophoric canal.* In male the two lateral borders of its body are curved (on ventral sides) to form a groove the gyanophoric canal, in which the female lodges.

Internal structures

Digestive system. Anteriorly-placed mouth surrounded by oral sucker, leading into a small oesophagus. The pharynx is absent. The oesophagus leads into a forked intestine which again fused posteriorly to form single structure for a distance. The intestine ends blindly. The oesophagus is surrounded by clusters of oesophageal glands.

Excretory system. It is represented by a bladder and a pair of collecting tubules. The bladder opens out through a single minute excretory pore lying ventrally near the acetabulum. The flame cells collect waste products from mesenchyme and discharge into collecting tubules to the bladder.

Nervous system. It is similar to other trematodes consisting of a circum oral ring with coerebral ganglia and two lateral longitudinal nerve cords.

Reproductive system.

Sexes are separate. Males are broader and enclose the females in the gynecophoric canals.

1. *Male Reproductive system.* The male reproductive system consists of 4-7 *testes* which are spherical and are situated anteriorly. Each testis gives rise a minute duct known as vas efferens. All the *vasa efferentia* unite to form a common spermduct or *vas deferens*, which after a short distance dilates into a structure, the *seminal vesicle* just before opens into a genital aperture situated just below the acetabulum or ventral sucker.

2. *Female reproductive system.* Female reproductive system consists of a single ovary. From the posterior portion of *ovary* arises a coiled tube known as *oviduct.* The oviduct bends upwards and runs after a short distance to open into a ootype. The ootype in turn gives off a wide coiled tube, the *uterus*. The uterus is thin and long tube running between the two branches of intestine and opens into a female genital pore near the acetabulum. The *ootype* is surrounded by Mehli's' gland. *Vitelline glands* are situated in the posterior region of the body. From these glands a vitelline duct arises that runs along the oviduct and joins the latter before it opens into ootype. In *S. haematobium* seminal recepticles is absent.

Life-history

Copulation and fertilization. The male and female adults in the definitive host move against the blood flow into small venules until they can move no further. Fertilization occurs when spermatozoa from the male are introduced into the female. Fertilization takes place in the oriduct. The fertilized eggs are encapsulated by the secretion of vitelline glands which also form yolk cells. In *S. haematobium* the uterus contains only 10-100 capsules while uterus of female *S. mansoni* contains only 14 capsules at a time.

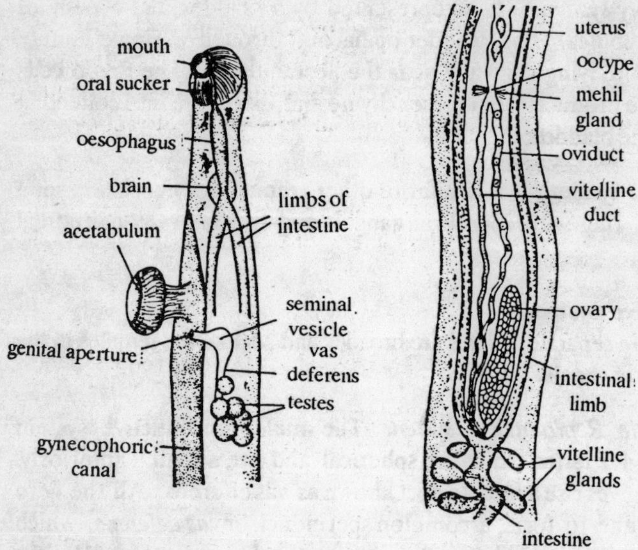

Fig. 4.2. A = Antomy of male *S. mansoni*. B = Anatomy of female *S. haematobium*.

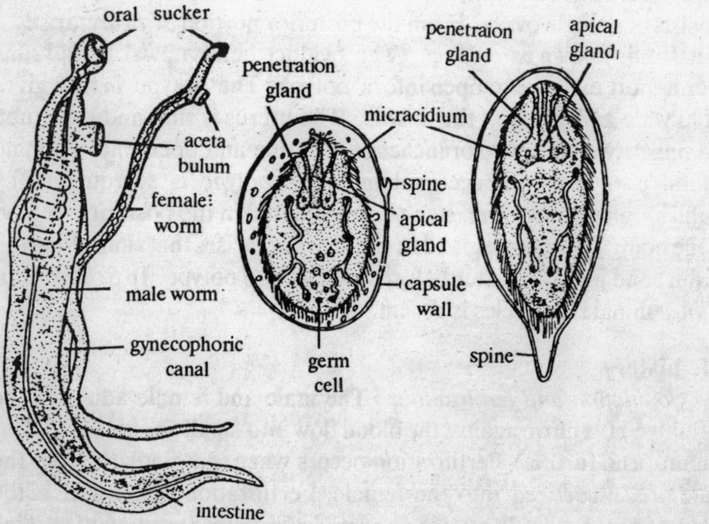

Fig. 4.3. A = *Schistosoma japonicum*. B = Capsule of *S. japonicum*. C = Capsule of *S. haematobium*.

Egg laying. The fully formed eggs or capsules measures 114-175 m x 45-68 u in *S. mansoni*. In *S. japonicum* 70-100 u x 50-65 u and in *S. haematobium* 112-170 u x 40-70 u. The capsules are nonoperculate and bear a large posterio-lateral spine in S. mansoni. In *S. haematobium*, the spine is terminal and in *S. japonicum* the spine is small and subterminal. In *S. mansoni* such eggs are laid in the mesenteric vessels and are carried by blood to the proximity of large intestine where they gradually penetrate the intestinal wall, become included in the lumen, and are passed out in feces. In *S. haematobium* the capsules are passed into the urinary bladder and are expelled in urine.

Female lays a single egg at a time. *Pesigan et al.* (1958) estimated that a female *S. japonicum* lays about 1200 eggs per day.

Miracidium larva. Eggs do not hatch in the urinary bladder or intestine of the host. The eggs of *S. japonicum* takes 9-10 days to develop mature miracidia while passing through the tissue and will live for 10 or 12 days longer if not expelled out, but may die in the tissue. Miracidium measures about 130 u in length and 60 u in breadth. The larva contains ciliated epithelium. Two pairs of penetration glands are present, one pair anterior in position other posterior. Meracidium also possesses a primitive gut, nerve centre and two pairs of flame cells. Germ balls are usually located in the posterior portion of the body.

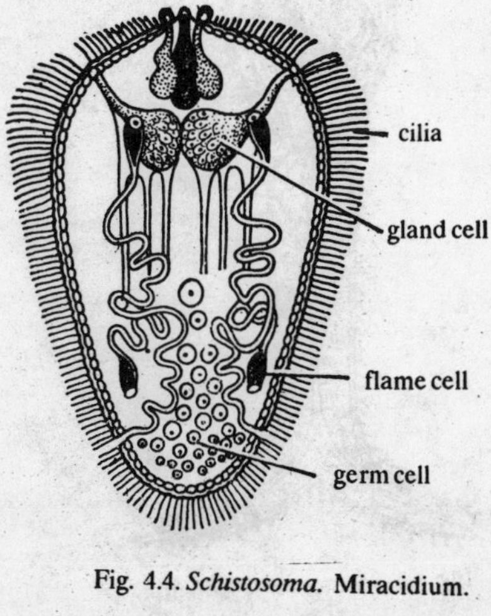

Fig. 4.4. *Schistosoma.* Miracidium.

When the eggs come in contact with water, at 28°C the enclosed miracidia break out the shells by cracking them. The free-swimming miracidia finds and penetrates a pulmonate snail and transforms into mother sporocyst which is located in the foot region of the host. Free-swimming miracidium survive for upto 24 hours. It dies if not find its intermediate host within this single day of free life.

Sporocyst. In *S. mansoni* the sporocyst grows 1 mm in about 14 days. The sporocysts start producing daughter sporocysts which burst free of the mother sporocyst. These are differentiated from germ balls. The daughter sporocysts reach a length of 1.5 mm and produce forktailed cercaria from the germ balls. A single snail infected by a single miracidium may discharge about 3500 cercariae per day for sometime. The cercariae are discharged during the forenoon on sunny days.

Cercaria larva. The cercariae of *S. haematobium* and *S. mansoni* have a body of about 200 M long with a tail stem slightly longer and forks about 75 M long. For about two or three days the cercariae alternately swim and rest in water. Cercaria possesses an oral sucker and a ventral

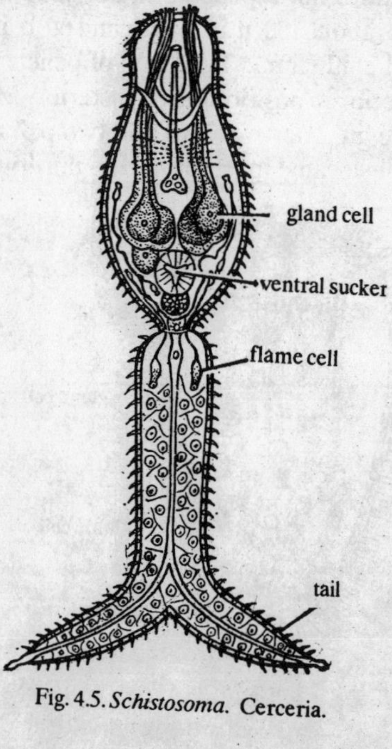

Fig. 4.5. *Schistosoma.* Cerceria.

acetabulum are present. Mouth leads into the oesophagus which leads into a trifid intestine. Pharynx is absent. Five pairs of penetration glands are present - 2 pairs of anterior oxyphilic glands and 3 pairs of posterior basophilic glands. If the cercariae fail to reach final host within two days they die.

Infection to the final host.

If successful the cercariae burrow through the skin, using the histolytic and hyaluronidase - bearing secretion of their penetration glands just as the miracidia do. Skin penetration requires several minutes and may or may not be accompanied by a prickling sensation and subsequent dermatitis. If ingested with water the cercariae attach themselves to the mucous membranes of the mouth or throat and similarly bore in. They leave their tail behind them. They find their way into the blood system and are carried via the heart to the lungs. Young *S. mansoni* accumulate in the lungs on the 2nd and 3rd days. By the 6th day they appear in number in the liver, where they are well established by the 15th day. Apparently cercariae feed only on portal blood, but once in the liver they grow rapidly. Migration to the mesenteric veins begins about the 23rd day, and mating and egg production about the 40th day. All the larval produced from one egg become adults of some sex. No female worms mature in a host, if the male worms are absent.

Pathogenicity

The injuries inflicted by schistosomes depend on the degree of infection and the prolonged period of infection and reinfection. During migration of the eggs through tissue to reach the colon or urinary bladder tissue reactions in the form of inflammation, necrosis, connective tissue encapsulation, eventually scar formation take place. During migration of the eggs through the wall of urinary bladder or colon, haemorrhage generally occurs. This is accompanied by diarrhoea.

In endemic areas, where reinfection occur, repeated penetration of colonal and sladder wall results in extensive scar formation, which prevents migration of eggs through these structures. As a result, the eggs may transported to liver and rarely to brain or spinal cord. Whenever these aberrant eggs become lodged, local inflammation, fibrosis and necrosis occur. Cirrhosis of the liver is not uncommon. Swelling of liver and spleen is of common occurrence. In the case of heavy infections, mature worms may migrate to the brain, lungs, uterus, oviduct and gonads. Their presence at these sites provoke pathological changes.

Acute cases of schistosomiasis, especially by *S. japonicum*, which is highly pathogenic, often lead to death.

Treatment

Treatment of schistosomiasis has been hundred by the fact that relatively few drugs are available, their toxic effects are sometimes severe and they are active only against mature worms. Tartar emetic (potassium antimony tartrate), Sodium antimonial gluconate and Stibocaptate have been used for years. Thioxanthones, miracils, Lucanthone, hycanthone, Niridazole imidazolidin z-one, etc. are also used.

Epidemiology and Control

The basic pattern of life cycles in the blood flukes was worked out in the first decade of the 20th century. With that information came the keys to control. It was obvious that sanitary disposal of human excreta and reducing population of snails were the steps to take. As knowledge increased, additional techniques were developed, so that general methods used in attemping to control schistosomiasis are :

1. Snail control
2. Sanitary disposal of human excreta
3. Treatment of infected persons
4. Treatment or elemination of reservoir host.
5. Prevention of penetration by cercariae.
6. Immunization of persons at risk of infection.
7. Prevention of infections in snails.

Each method is directed towards the same end ; to break the life cycle or interrupt the chain of infection. It all the snails could be removed from endemic area, the probablity of infection would approach zero. If all definitive hosts were treated and maintained free of infection for a year, the cycle would be broken.

Table Showing differential characteristics of *Schistosoma* species of humans

	S. haematobium	*S. japonicum*	*S. mansoni*
Adults			
Male			
Length(mm)	10-14	12-20	6-12
No.of tesets	4	7	6-9
Tegument	scattered tubercles and blunt short spines	Minute spines 1	Numerous tubercules and large sharp spines.
Posterior union of intestinal caeca midbody.	Midbody	Posterior to	Anterior to midbody
Female			
Length (mm)	16-20	16-28	7-17
ovary	Posterior to midbody	Midbody	Anterior to midbody
No.of eggs in uterus	20-100	50 or more	usually 1
Eggs produced by single female per day.	?	1400-3500	100-500
Location in host			
Primary host	Bladder and pelvic plexuses	Both superior and inferior mesenteric veins	Veins of small and large intestine near ileocecal junction.
secondary host	Portal and mesenteric veins	Portal vein	Haemorrhoidal and hepatic veins
Prepatant period in weeks.	10-12	5-7	7-8

Eggs			
Length(u)	112-170	70-100	114-175
Mean		82	
Width(u)	40-70	50-65	45-68
mean		60	
Colour	Golden brown	Golden brown	Golden brown
Spines	Terminal	Subterminal knob	Subterminal spine
Passed in	urine	faeces	faeces

5

OPISTHORCHIS SINENSIS

Opisthorchis sinensis is commonly known as 'Chinese liver-fluke'. It was formarly known as *clonorchis sinensis* and was described in 1875 from a chinese Hospital patient whose liver, on post-mortem, was found to be enlarged and turgid and from whose distended bile ducts many small trematode worms were recovered. Since than its epidemiology, therapy, morphology, prevention and life-cycle have been studied in details.

SYSTEMATIC POSITION

Phylum	-	**Platyhelminthes**
Class	-	**Trematoda**
Order	-	**Digenea**
Family	-	**Opisthorchidae**
Genus	-	***Opisthorchis***
Species	-	***sinensis***

Distribution

Opisthorchis is common in Japan, Korea, China, Hong Kong and Indochina. This distribution coincides with the distribution of the snail first intermediate host, *Parafossarulus manchouricus* and closely related hydrorbid snails (*Malek*, 1980). Nearly 100% of people in certain villages of China (*Komiya*, 1966) and over 65% of population of Hong Kong are infected.

Host and Location

Humans are definitive host for *Opisthorchis*. Most fish-eating mammals and fish-eating birds can be infected, but dogs and cats are probably the most important reservoir hosts.

Morphology

Opisthorchis is a smaller and narrower than *Fasciola* and not commonly related to it. Adult flukes are 10-25 mm in length and 2-4 mm in width. The body remains covered by rough cuticle. There are two suckers an oral and a ventral or acetabulum. The acetabulum is slightly smaller than the oral sucker and is located near the end of the anterior third of the body. Genital pore opens ventrally just infront of the acetabulum. An excretory pore lies at the posterior end.

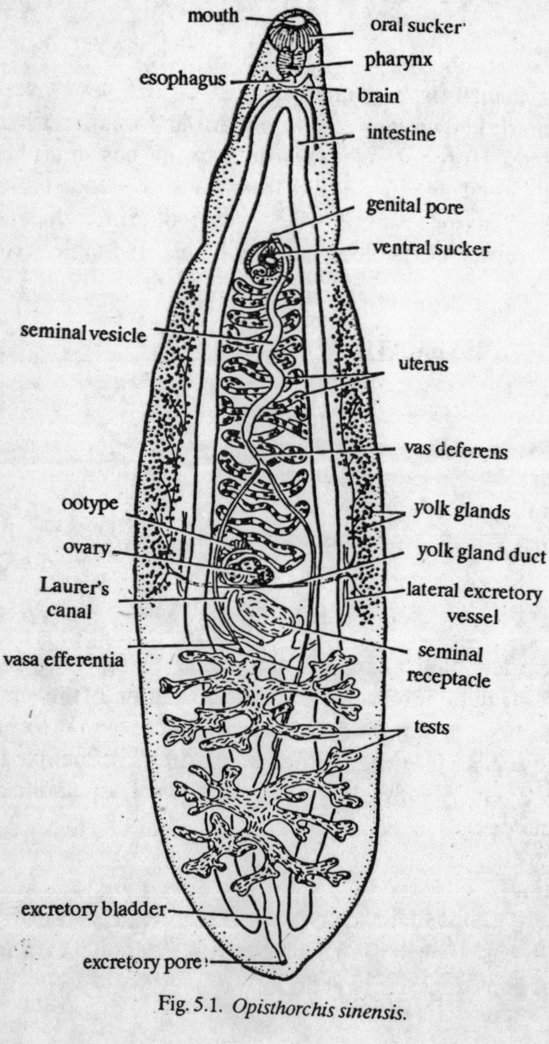

Fig. 5.1. *Opisthorchis sinensis.*

Integument and muscular system

A thin, minutely wrinkled cuticula which is devoid of spines overlies the hypodermis and the peripheral muscular layers of the body.

Parenchyma. The space between the organ system in the interior of the body is filled with a mesenchyme called parenchyma. This tissue consists of irregularly shaped cells between which is a matrix of fluid and intertwining fibres.

Digestive System

A transversely oral *mouth* which is situated subterminally opens into buccal cavity. It is surrounded by well-defined oral sucker occupying most of the space of the extremity. Posterior to the oral sucker is a conspicuous ovoid *Pharynx*. The sucker and pharynx which are provided with radiating muscle fibres, serve for attachment, abrasion of the bile ducts and aspiration of loosened food particles. Pharynx opens into a short *oesophagus* approximately equal to the size of pharynx. Oesophagus joins two mildly distended gut caeca. The buccal cavity, pharynx and oesophagus are lined with thin cuticula while the gut caeca, which reach nearly to the posterior end of the body, are lined with a vacuolated columnar epithelium. The epithelial cells secrete enzymes. The food consists of blood corpuscles and a mucin-like exudate produced by laceration of the inflamed mucosa of bite duct.

Respiratory system

The respiratory organs are absent and the animal is anaerobic because of low oxygen contents in its surroundings. Energy in obtained by anaerobic fermentation, the carbon dioxide, a by-product, is diffused out through general body surface in the surrounding medium.

Excretory system

A clear, irregular *excretory bladder* is constricted posteriorly to form a narrow efferent canal which opens to the outside by a posterior *excretory pore.* The bladder proper is approximately has the same diameter as that of a gut caecum. It follows anteriorly between the main bodies of the testes to the level of the seminal receptacle.

Two lateral *excretory vessels* join the bladder slightly behind its blind anterior end. These vessels curve laterally and forward and follow an extracaecal course to the level of the acetabulum, where they branch into anterior and posterior recurving tributary vessels. The continued branching of these major vessels results in increasingly finer vessels, the terminal units of which are the flame cells.

Nervous system

A bilobed brain with medial commissure and anterior and posterior nerves is located in the dorsal aesophaged region.

Reproductive system

Opisthorchis is hermaphrodite trematode possesses well-defined reproductive organs and a common genital pore.

Male reproductive system

Testes are posterior, branched and lying one behind the other. From the main body of each testis arises a thin duct the *vas efferens*. This duct proceeds forward to a point slightly anterior to the ovary where it joins its mate from the other testis to form a vas deferens vas deferens enlarges into a sperm-filled seminal vesicle. A narrow ejaculatory duct connects the vesicle to the genital atrium which opens at the ventral genital pore. A cirrus sac, prostate and cirrus all are absent.

Female reproductive system

A singe branched ovary is present in the median line slightly anterior to an oblong sperm-filled seminal recepticle. The *seminal recepticle* lies just ahead of the anterior testis. Anteriorly the seminal recepticle gives off a short duct that joins Laurer's canal and which terminally joins a short *oviduct. Laurer's canal* curves posteriorly around the seminal recepticle and also runs abruptly dorsally to open by a small aperture, the *opening of laurer's canal.* This opening lies between the anterior testis and seminal recepticle near the middle. The oviduct, after receiving the seminal recepticle, proceeds forward a short distance and receives a *common vitelline duct.* Vitelline glands are small and numerous located in the middle third of the body. Two lateral vitelline duct arises from the middle of transverse vitelline duct. The oviduct receiving all these ducts (i.e. median vitelline duct, Laurer's canal and seminal recepticle) then penetrates an *ootype* which is surrounded by *Mehlis'* gland. The Mehlis' glands are loosely organized and composed of small acinus shaped cells. A very conspicuous uterus arises from the ootype anteriorly and proceeds an on irregularly coiled duct to the genital atrium. The uterus becomes larger in diameter anteriorly because of the massing of eggs.

Fertilization

Mutual cross- fertilization occurs after copulation between two flukes. The ova liberated from ovary are fertilized by the sperms from the seminal recepticle. In the ootype the fertilized egg yolk are enclosed in a thin walled, oval, operculate shall which is often provided with a comma-shaped posterior appendage. The egg gradually acquires a darker brown colour and proceeds with cleavage until a fully formed miracidium is formed. The embryonated capsules, pass to out side through the genital pore and are carried by the bile to the host's intestine, they are finally eleminated with the faces.

Life-history

Man is the primary host besides snail and fish are two intermediate hosts.

Miracidium and sporocyst

As stated above the eggs when laid contain fully mature miracidia. Each ovoid miracidium bears long, motile cilia on an outer ectodermal sheath. The apical region bearing no cilia. In the apical region there is no apical spine and an apical gland opening. On the dorsal side there is a single excretory pore. An apical histolytic gland and a large, ventral, sausage-shaped secretory gland with a narrow duct opening besides the spine. About 25 germ balls are present, that fills the miracidial body space. Two dorsally placed flame bulbs and their ducts comprise the excretory system. Nervous system is situated ventral to the apical gland.

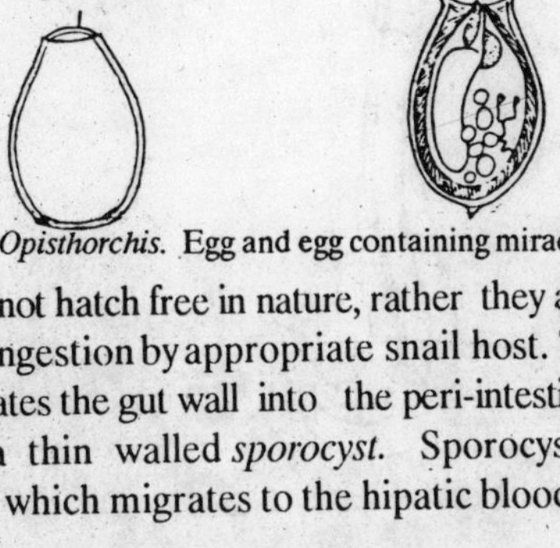

Fig. 5.2. *Opisthorchis.* Egg and egg containing miracidium.

Miracidium do not hatch free in nature, rather they are released from the eggs only after ingestion by appropriate snail host. The miracidium than actively penetrates the gut wall into the peri-intestinal blood sinus. Now it becomes a thin walled *sporocyst.* Sporocyst produces the redial generation which migrates to the hipatic blood sinuses.

Redia. A mature redia is elongated in appearance. Its anterior end is rounded whereas posterior end is some what pointed. The larva internally lined with sub-epithelium. A birth pore is absent. An anterior pharynx and a brownish sacculate gut extending about half-way posteriorly are present. Several germ balls are present which develop synchronously into cercariae. About 6-8 cercariae are liberated by the ruptuse of redial wall.

Cercaria. The long-tailed, oculated cercaria emerges from the snail host into the water and swims vigrously in the search of another host i.e. a fish. The tail is provided with muscle-fibres and bears thin, lateral, transparent elytra. Two bean-shaped eye-spots are located near the anterior end of body. An oral-sucker, pharynx and acetabulum are present but oesophagus and gut caeca are not formed yet. Seven pairs of unicellular penetration glands occupy. The middle of the body. Their ducts open near the mouth. Dorsal and ventral peripheral mesenchymal cells are packed with cystogenous glands. These glands

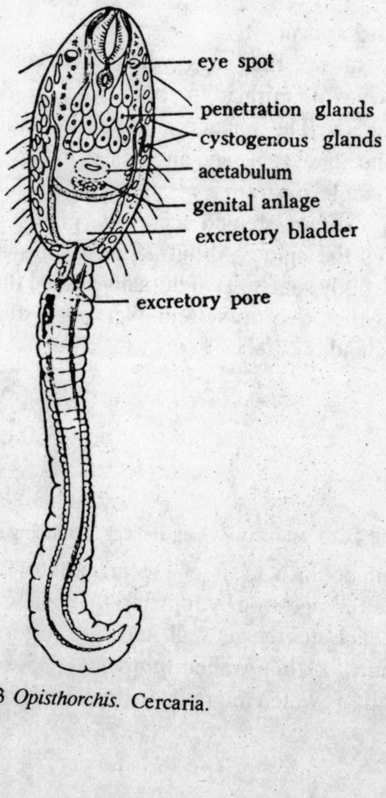

Fig. 5.3 *Opisthorchis.* Cercaria.

help in the formation of cyst of metacercaria. Excretory system comprises a triangular excretory bladder with anterio-lateral collecting ducts and a bifurcated posterior caudal excretory tubule.

Metacercaria. On coming into contact with the fish, member of Cyprinidae family, the cercaria attaches to the surface, discards its tail, and creeps under a scale into the fish. A cyst wall is formed by the material secreted by cystogenous glands. The cyst is thin, tough, and spherical. Organ formation continues e.g., the gut caeca etc.

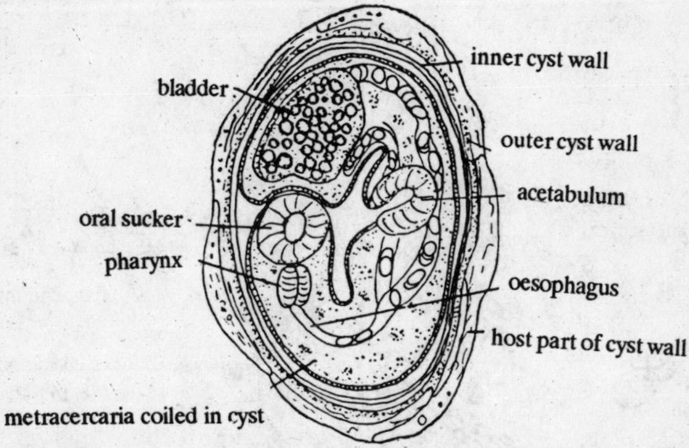

Fig. 5.4. *Opisthorchis.* Metacercaria

Inspection to the primary host. Primary host i.e. man ingests uncooked fish containing metacercaria. Metacercarial cyst walls dissolve in the duodenum or intestine (*Gibson* and *Sun,* 1971) and juvenile flukes migrate up the common bile-duct into the small bile-ducts of liver (*Sun et al,* 1968). Egg laying begins 3-4 weeks after infection.

Effects on the host

Symptoms are classified as mild, modrate or severe, depending upon the burden of worm, length of infection and health of host. Light infections may be asymptomatic. Heavy infections are characterized by chills and fever, followed by liver enlargement and tenderness, sometimes with splenomegaly (Spleen enlargement). Chronic heavy infections cause cholecystitis (inflammation of gall bladder) and hepatitis (inflammation of liver). Modrate infection causes diarrhoea, edema and some hepatomegaly (enlargement of liver). The disease is termed *opisthorchiasis* or *clonorchiasis.*

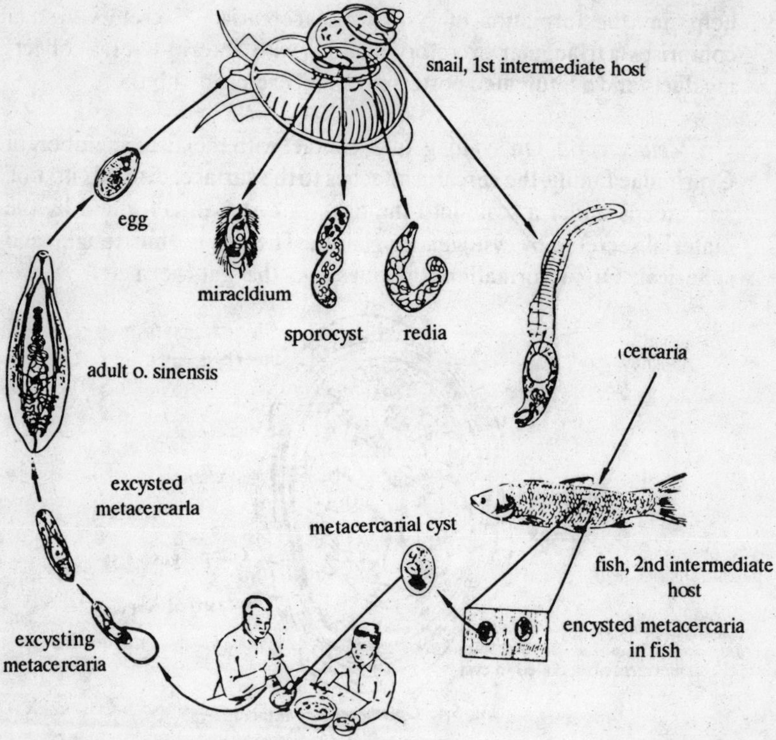

Fig. 5.5 *Opisthorchis.* Life cycle.

Treatment

Hexachloroparaxylol, Hetol and Paraziquantal are used. The first two appear to be useful.

Control

All the fishes, used as food, should be properly cooked to avoid infection.

Molluscicides to control the snail (first intermediate host) are generally not used because they are harmful to important fresh water fishes *(Malek*, 1980). Animal and human faces are important fertilizers for fish ponds in the Orient. Faeces must be properly composted. In addition, snail populations have been significantly reduced by removal of water plants when snails lay their eggs, thereby permitting fish more easily to feed on larval snails.

6

TAENIA SOLIUM

Taenia solium belongs to class Cestoda. The members of this class are commonly known as 'Tapeworms'. All are endoparasites in the alimentary canal and associated ducts of various vertebrate hosts. During their life cycles one or two or more intermediate hosts, vertebrate and invertebrate, are required in which the tapeworms undergo a phase of development. The cestodes differ, however, from the members of other two classes (Turbellaira and Trematoda) in the complete absence of a digestive tract.

The present text mainly related to *Taenia solium,* the pork-tapeworm of man.

SYSTEMATIC POSITION

Phylum	-	**Platyhelminthes**
Class	-	**Cestoda**
Subclass	-	**Eucestoda**
Order	-	**Cyclophyllidea**
Genus	-	***Taenia***
Species	-	***solium***

Habits and habitat

Taena solium or pork tapeworm is commonly found in those parts of the world where pork is eaten as food without properly cooking. It is found as an endoparasite in the gut of vertebrate hosts. One host is man and another is pig. Formerly, it was very common intestinal parasite in Europeans, especially in some places of Europe. It is rare in many parts of the world *viz.,* India, Philippines, North Amercia. It is not at all found in Jewish and Mohammedan countries as in those countries eating of the flesh of pig is considered as an evil deed.

External morphology

Shape, size and colour. The body is elongate and flatenned dorsoventrally like a ribbon or tape. The body is narrow anteriorly and gradually broadens laterally towards posterior end.

The size of adult worm varies from 3-5 metres but a few are recorded to attain a length of about 8 meters.

The body is usually opaque white in colour but creamish, yellowish or greyish colouration is also common.

Division of body

The body of *Taenia* can be divided into three regions.

1. Scolex
2. Neck
3. Strobila

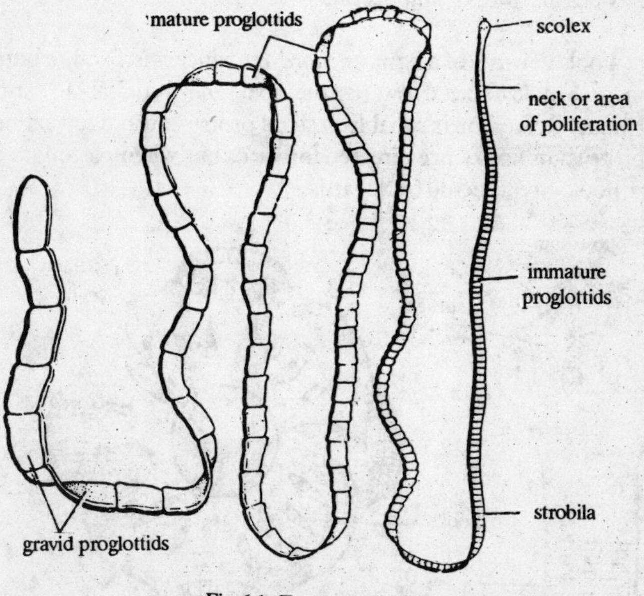

Fig. 6.1. *Taenio solium.* Entire.

1. *Scolex.* The scolex is the holdfast organ, which is generally considered the anterior end. There are two schools of thought concerning the interpretation of the structure of tapeworms. Some consider a single scolex and connected proglottids to be a single animal. Others consider each proglottid to be an individual - hence, a single tapeworm is a colony of many individuals. Some helminthologists argue that the scolical end of the cestode is actually the posterior end. However, *Hyman* (1951) and *Chang* (1964) are not convinced with the argue that scolex is the posterior end. Furthermore, the more recent literature indicates the scolocal end should be considered the anterior end.

Scolex is a pear-shaped structure, but is four-sided at the posterior part. It possesses a small terminal cone called the *roseteltum*. Around the base of the terminal cone or rostellum is a double row of curved, pointed, chitinous hooks, which are usually 28 in number but are sometimes between 22 to 32. Hooks are of two types, long and small. They are arranged alternately. Long hooks are approximately of 180u and short ones are of 130 dimension.

Each hook consists of a *base* or *guard* by which it is fixed, a blunt projection or handle directed towards the apex and a conical outwardly directed *blade.* Rostellum is capable of slight protrusion and retraction. During protrusion hooks are directed forwards but when rostellum is retracted, hooks are directed backwards.

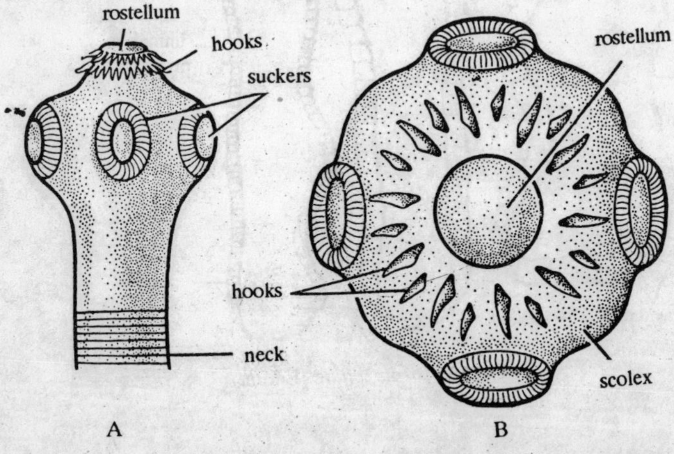

Fig. 6.2. *Taenia solium.* A = Scolex in magnified, B = Frontal view

The mouth is absent in scolex. The term "head" frequently used for the scolex is, thus, inappropriate, because the head is essentially that part of the body where organs for *perceiving* and catching food are aggregated. It measures about 0.6-1 mm in diameter.

2. *Neck.* It is situated immediately posterior to the scolex. This is an unsegmented, poorly differentiated area that is generally narrower than the scolex and the strobila proper, and is the continuously differentiating or budding zone or area of proliferation that gives rise to immature proglottids.

3. *Strobila.* The strobila, which conatitutes the main mulk of body and is made up of a chain of proglottids. There are about 800-1000 such *proglottids*. Each proglottid bears a complete set of genitalia. This linear repetition of genital organs is termed as *proglottisation*. Since the proglottids are budded off from the neck region in an orderly succession, the youngset proglottids are towards the neck and the oldest proglottids are posterior most.

Depending upon the development of genital organs, the proglottids are distinguished into three types :-

(i) *Immature proglottids* - These are young, undifferentiated proglottids neck to the next region. The reproductive organs are either absent or in stages of development. Their width is more than the length. They comprises about 200 anterior proglottids.

(ii) *Mature or reproductive proglottids* - The mature proglottids are sexually mature consisting of a male and a female set of reproductive organs. They are usually as long as broad and are found in the middle of the body. They comprise about 450 proglottids *Taenia* is protandrous, about 100 or 150 anterior mature proglottids contain only male reproductive organs and remaining 300-350 contain both, male and female reproductive, organs.

A mature proglottid is a complete reproductive unit in its own right and may underago self or cross fertilization.

(iii) *Gravid or ripe proglottid* - These are found in the posterior part of the body. These are characterised by the possession of a large, highly branched uterus filled with developing embryos. The rest of the genital organs have already disappeared from them after performing their function. They are longer than broad. The posterior most proglottids are cut off from the body and are expelled out of houst body along with its faeces. This process of separation of proglottids is known as *apolysis.*

Body-wall

The body wall of cestodes is made up of several layers but lacks a cellular epidermis. It consists of (i) cuticle (ii) Basement membrane and (iii) Musculature.

(i) *Cuticle.* The outer most layer of the body surface is called the cuticle or tegument. This layer is more or less transparent, noncellular, and without visible pores. The cuticle consists of two layers. (a) outermost hair-like or fringe-like *comidial layer* and (b) a thick *homogenous* layer. The cuticle is secreted by modified mesenchymel cells. These cells lie beneath the musculature and have long processes extanding up to the cuticle.

When studied under electron microscope, minute projections known as *microtriches* are present on the cuticular surface. *Rothman* (1963) suggested that the distal portion of each microthrix, which is solid, is concerned with two functions (a) it may serve as a means of resisting the intestinal current, since the body surface always in contact with the microvilli of the straited border of the cell lining of small intestine of host (b) it may serve to agitate the microhabitat in the vicinity as the worm moves, thus stirring up the intestinal fluids so that mutrient materials as well as waste products are always in a state of flux. The proximal postion of microthrix is medullated and could very well serve as sites of absorption. In addition to microtriches, ultramicroscopic pore canals that communicate with the parenchyma are present in cuticle. These may be the sites of absorption also. Mitochondria and endoplasmic canals are also reported in the cuticle. Thus, it is apparent that this outer body layer is a dynamic structure rather than a mere covering.

(ii) *Basement membrane.* Immediately benath the cuticle is the basement ment membrane. This noncellular connective tissue layer contains vacuoles and granules. The protoplasmine layer beneath the basement layer is thick.

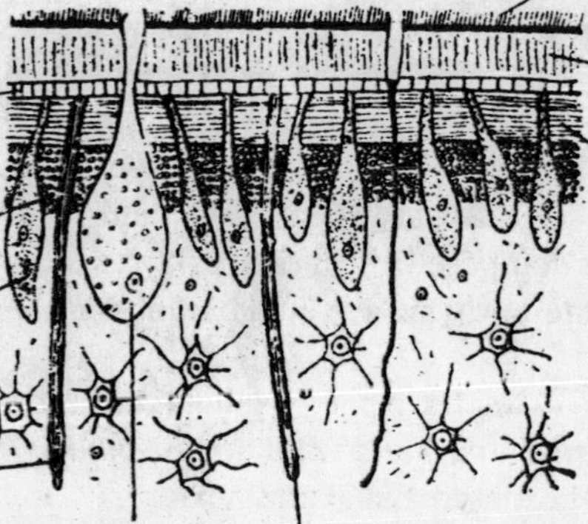

Fig. 6.3. *Taenia solium.* V.L.S. body wall seen under light microscope.

(iii) *Musculature.* Immediately beneath the basement membrane is a layer of *circular muscles* followed by *longitudinal muscles* and vertical or *dorsoventral museles.*

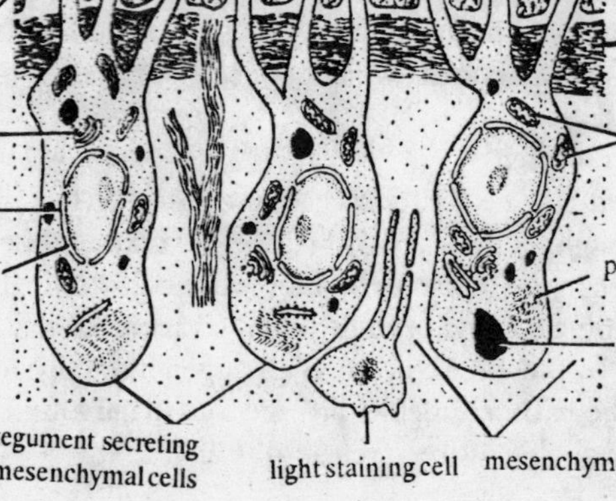

Fig. 6.4. *Taenia solium.* V.L.S. Body wall seen under electron microscope

Mesenchyma or parenchyma

The space enclosed by the body wall, except for the occupied by the reproductive organs, osmoreguletroy structures, muscle fibres and nervous tissue, is filled with a spongy type of tissue, the parenchyma. Certain syncytially arranged cells form a network. In living animals, the space between the cells are filled with fluid. Numerous rounded calcarious bodies composed of concentric layers of calcium carbonate are present in mesenchyma secreted by *lime cells*.

Except at the two sides in each segment the parenchyma is divided into (i) outer *cortical* past and (ii) inner *meduallary* part by ciruclar muscle fibres of the mesenctymal musculature.

The parenchymal cells, and to some extent the spaces, serve as sites for the storage of glycogen.

Nutrition

In *Taenia* there is no alimentory canal and as it lives in the intestine of host, there is totally absence of alimentary canal in all the stages of its life-cycle.

Its chief source of nourishment is obtained from the intestine of host. These are amino acids, glucose, glycerol etc.

Chandler thought that tape-worm gets its sugar supply from digeated food-material of a host but *proteins* and *vitamins* are made available to the animal from the living tissue of the host by the secretion of the digestive enzymes, which effect the cells.

Respiratory system

In *Taenia*, the respiratory organs are absent so that respiration takes place through general body surface. Like other endoparasites, *Taenia* lives in the intestine of man where the oxygen content is very low or absent. So mode of respiration is anaerobic in which the energy is obtained by the break-down of glycogen into fally acids and carbon dioxide. The carbon dioxide diffused out in surrounding medium. If oxygen is available, it respires aerobically.

Excretory system

Excretory system, also known as osmoregulatroy system, is same as the protonephric type found in tremetodes.

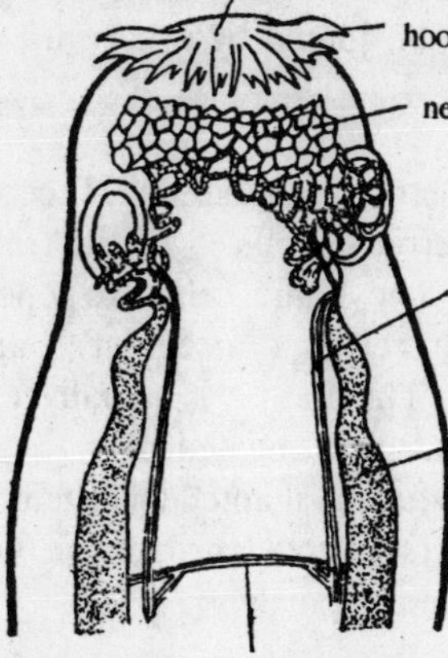

Fig. 6.5. *Taenia solium.* Nephridial plexus.

Four main collecting canals traverse the entire length of the strobila. The two ventral canals are ventrolaterally located, and the two dorsal canals are dorsolateral. All these canals are situated in the peripharynageal zone of the meduallary parenchyma. A single transverse canal connects the two ventral canals at the posterior end of each proglottid. The ventral vessel carry the water away from the

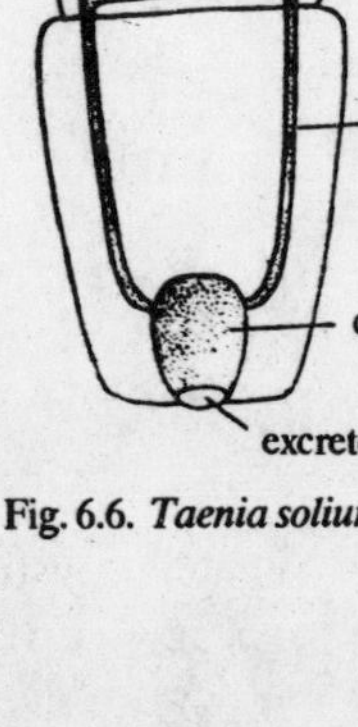

Fig. 6.6. *Taenia solium.* Caudal vesicle in last segment.

scolex, the dorsal vessels toward it. Within the scolex, the four longitudinal canals may be joined by a network of canals forming *nephridial plexus.*

Along the length of ventral canals, a series of secondary tubules arise that may in give rise to tertiary tubules. At the free end of terminal tubules are flame cells. The flame cells are generally arranged in a group of four. A flame cell is of irregular shape, with granular cytoplasm and nucleus. The 'flame' is actually a group of cilia that arise from a concave basal plate located near the cell nucleus. The cilia are enveloped in the funnel-shapted enlargement of the free end of the tubule waste products collected through the flame cells is passed down the tubules into the main canals.

In the young worm, there is an *excretory vesicle* in the posterior most broglottid, into which the ventral canals open. However, as the neck region produces more proglottids, the older proglottids are pushed further back, and eventually the proglottid containing excretory vesicle is detached by the process of apolisis. Thus, in most older animal the posterior end of the tubes open independently to the exterior.

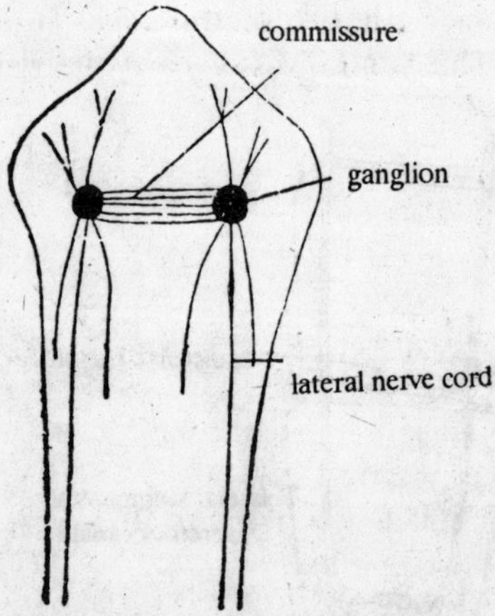

Fig. 6.7. *Taenia solium.* Nervous system.

Nervous system

A thick band in the scolex with two ganglia constitute the brain or central nervous system. From the brain few nerves are given out anteriorly which innervate the scolex. Three pairs of longer nerves are given out towards the posterior. Out of these only one pair, the *lateral nerve, cords* extend through the strobila passing just outside the excretory canal in each proglottid.

Sense organs are neither present nor they are necessary. Free nerve-endings are found in the body wall and detached proglottids for some time show independent muscular movement.

Reproductive system

Taemia is *a hermaphrodite.* Reproductive organs are usually found in the proglottids after 200th partition. Some anterior proglottids are functionally male. Middle proglottids are both male as well as female. Proglottids at the posterior end are filled with enlarged *uterus*.

Structure of the proglottids in the middle region is as follows:-

I. The male *reproductive system* consist of the following organs:

1. Testes
2. Efferent ducts.
3. Vas deferens.
4. Cirrus or penis.
5. Cirrus sac.

1. *Testes.* They are numerous small, rounded bodies situated in the medullary past of the parenchyma. There number is about 15-200 and are distributed throughout close to the dorsal surface of the proglottid. Some workers are of the openion that the numerous rounded follicle, in fact, constitute a single and highly sub-divided testis. However, the number of testes is of texanomic value in tapeworms.

2. *Efferent ducts.* From each testis arises a fine ductule or capillary known as *vas efferens*. All the *vasa efferentia* unite to form a *vas deferens*.

3. *Vas deferens.* It is a long and tube, arrising from the mid-plant of the proglottid. It runs transversely to the left or right to open into cirrus. In some cestodes the vas deferens enlarge to form the seminal vesicle for the storage of sperms. But in *Taenia* no such enlargement is seen.

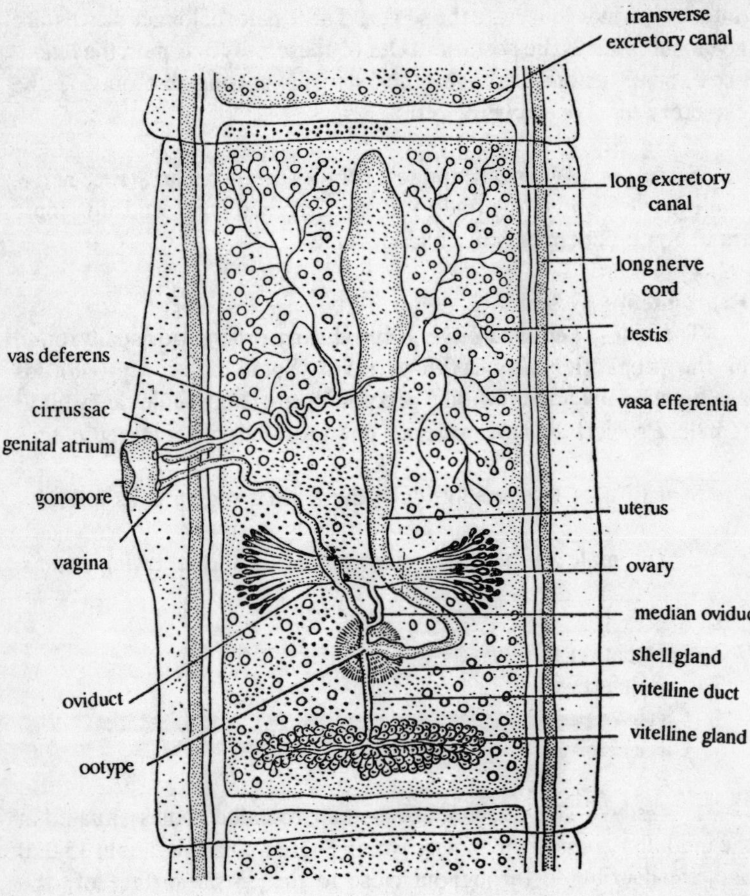

Fig. 6.8. *Taenia solum.* Mature proglottid showing reproductive system.

4. *Cirrus or penis.* The distal end of vas deferens in modified into a protusible *cirrus.* The cirrus is armed with bristles, hooks or spines. It is eversible through a male genital pore into a cup like *genital atrium* of the genital papilla.

5. *Cirrus sac.* Cirrus is surrounded by a kind of muscular sac known as cirrus sac. Its function is to erect penis during copulation.

II. *The female reproductive system* comprises :

1. Ovary
2. Oviduct
3. Ootype
4. Uterus
5. Vagina
6. Vitelline gland
7. Mehli's gland

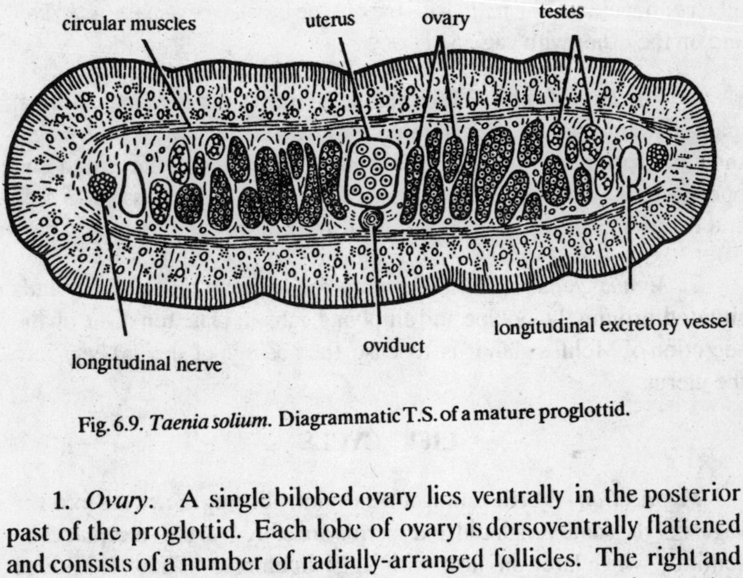

Fig. 6.9. *Taenia solium*. Diagrammatic T.S. of a mature proglottid.

1. *Ovary*. A single bilobed ovary lies ventrally in the posterior past of the proglottid. Each lobe of ovary is dorsoventrally flattened and consists of a number of radially-arranged follicles. The right and left lobes are connected medially by a tranaverse bridge, the *ovarian isthmus*.

2. *Oviduct* - The oviduct arises from the middle of the bridge. It is a short but wide duct and joins the ootype distally.

3. *Ootype* - It is a small rounded chamber developed at the junction of oviduct with the vitelline duct. It is surrounded by numerous unicellular *Mehlis glands*, the cells of which form an amorphous complex.

4. *Uterus* - A blind sac-like or tube-like uterus arises from the ootype and runs forward in the segment. In gravid proglottids the uterus enlarges and gets branched to occupy the whole of the proglottid. It remains filled with the fertilized ova or developing embryos.

5. *Vagina* - The narrow tubular vagina arises from the receptaculum seminis and runs obliquely outwards to open into the genital chamber by female reproductive opening.

The seminal receptacle or receptaculum seminis is a small wide tube connected on one hand with the ootype by a narrow *spermatic duct* and on the other with vagina.

6. *Vitelline gland* - The vitelline gland is a compact, elliptical mass of numerous follicles situated posterior to the ovaries. A single short median duct, *median vitelline duct* arise from the gland and opens into the ootype. The secretion of the vitelline duct is rich in yolk and forms a covering of yolk and shall around the fertilized egg.

7. *Mehlis glands* - These are numerous minute unicellular glands situated around the ootype and emptying into it. The function of the secretion of Mehli's gland is to ease the passage of shelled eggs into the uterus.

LIFE CYCLE

Fertilization - Fertilization occurs by introducing the cirrus into the vagina of the same proglottid and sperms are stored in the receptaculum seminis, which later on fuse with eggs in uterus. Thus in *Taenia* self-fertilization takes place. Sometimes, the anterior mature proglottids, devoid of female genitalia, can enter into copulation only with the posterior mature proglottids with fully developed female genital organs. Fertilization, following copulation between two different proglottids, is sometimes termed *cross-fertilization* to distinguish it from that occurring between gametes of the same proglottid (self-fertilization).

Thus the eggs are fertilized in the oviduct and get surrounded with yolk and egg-shall in the ootype. These materials are provided by vitelline gland. The capsulated eggs enter the uterus and are collected there. The uterus enlarges in size, gets branched and occupies the whole space. The eggs are small measuring 40μ (micron) in diameter.

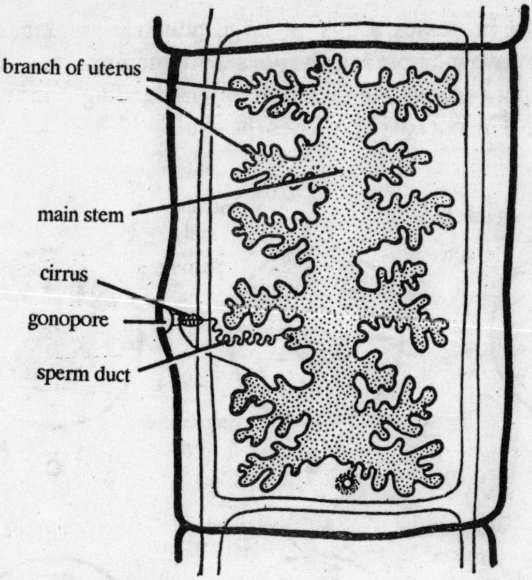

Fig. 6.10. *Taenia solium.* A gravid proglottid.

Cleavage - The zygote or egg cell undergoes cleavage when the capsule is in the uterus. The first unequal division results in a larger megamere and a smaller *embryonic cell*. The megamere divides further and froms several similar megameres, while the embyoonic cell divides repeatedly producing two types of embryonic cells, larger mesomeres and smaller micromeres. Thus, three types of cells result from the zygote; small micromeres, medium mesomeres and large megameres.

Onchosphere - By repeated division of micromeres a solid ball of cells, the *morula* is formed. The megameres divides, redivides and becomes filled up with yolk from the yolk cells, which surrounded the embryonic cells just like an envelope. It provides nourishment to the embryonic cells and finally disentigrates. The mesomeres form the inner thick striated membrane known as *embryophore*. Beneath embryophore lies basement membrane. The, morula, develops three pairs of chitinous hooks. These *hooks* are secreted by the differentiated cells known as *onchoblasts*. Now the six-hooked embryo is called *hexacanth.* It possesses a pair of large penetration glands and is surrounded by two membranes, the hexacanth membrances. The hexacanth, together with all the membranes surrounding it is known as *Onchosphere*. The onchosphere loses the shell.

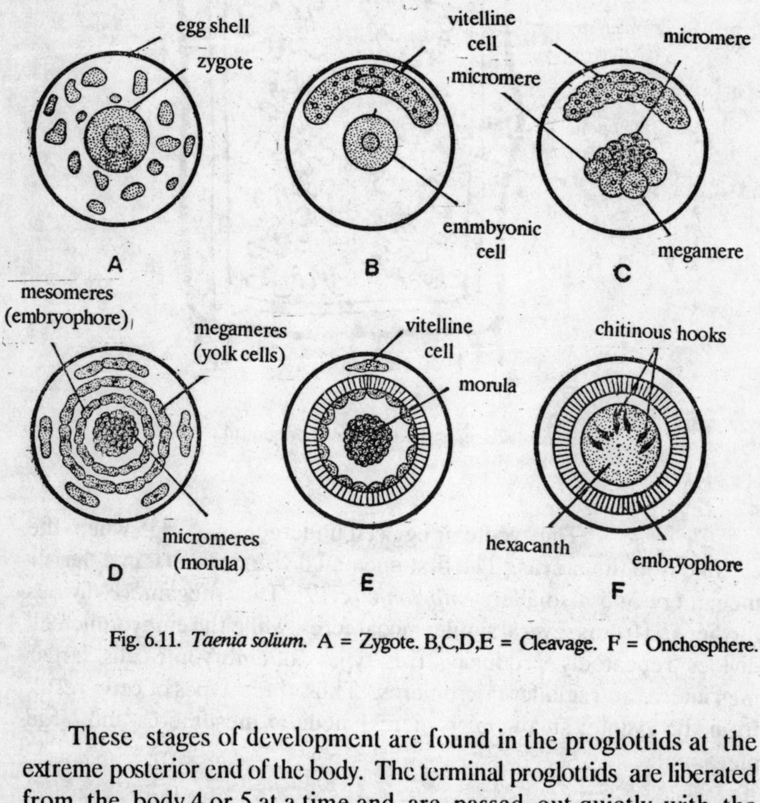

Fig. 6.11. *Taenia solium.* A = Zygote. B,C,D,E = Cleavage. F = Onchosphere.

These stages of development are found in the proglottids at the extreme posterior end of the body. The terminal proglottids are liberated from the body 4 or 5 at a time and are passed out quietly with the faeces of man. The proglottids show slow movements of contraction and later on perish. *Embryos* remain active for a considerable time. They are capable of further development only if they are taken in by a secondary host, which is a pig.

Infection to secondary host. The pig is the secondary host which regularly feeds of the faeces of man. Dogs, monkeys and sheeps are also known to get infection.

Inside the alimentary canal of a pig, proglottids and thousands of egg's shells, nearly 30,000 to 40,000, are digested and *hexacanths* are set free. Later on, hexacanths bore their way with the help of hooks and penetration glands, through the wall of the intestine and reach intestinal blood vessles. The process requires only 10 minutes.

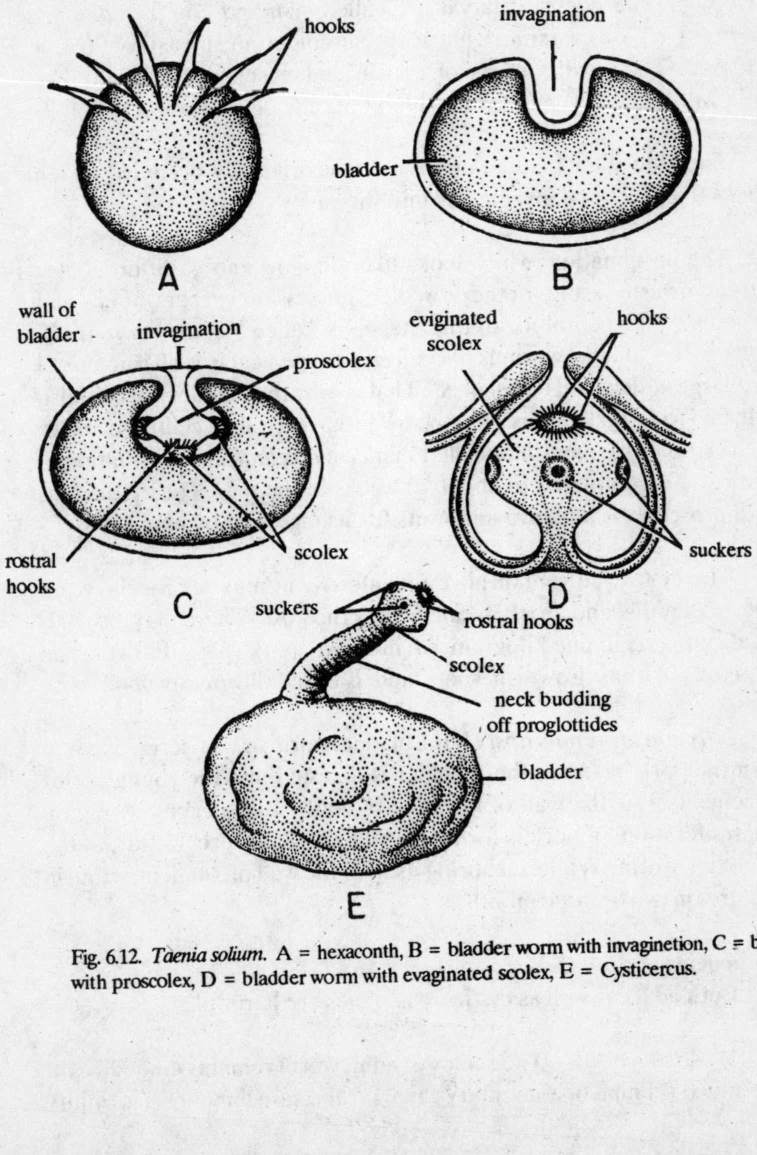

Fig. 6.12. *Taenia solium.* A = hexaconth, B = bladder worm with invaginetion, C = bladder with proscolex, D = bladder worm with evaginated scolex, E = Cysticercus.

Migration. Travelling through the heart, these enter the muscles of various parts in the body. The usual site where the hexacanths get encysted are the voluntary muscles of tongue, heart, liver and shoulder.

In the muscles, hexacanths lose their hooks, grow in size and develop a fluid-filled cavity in them. They now encyst in a cuticular covering. The encysted larval are called *cysticerci* or the *bladder worms.* They are oval somewhat whitish in colour and measuring 6 mm x 10 mm. Their wall consists of cuticle and mesenchyme layer. The fluid within them is mainly the blood plasme of the host.

The side of the bladder opposite to the originally occupied by the hooks thickness and invaginates into the cavity.

The invagination, which looks like a hollow knob, differentiates into an inverted scolex or the *proscolex*, possession suckers, hooks and rostellum. The embryo, at this stage, is called *bladderworm*. In *T. Solium* the bladderworm is of *cysticercus* type which is characterised by a large vesicle and one scolex. That is why the baldderworm is also referred to as cycticercus. It appears to have a wall of cellulose and is, for this reason, sometimes called *cysticercus cellulosae.* Formation of cysticerci is completed in about 10 weeks in the pig. The cysticercus with proscolex is infective and waits for its ingestion by man.

The cysticered remain alive and infective in muscles for 5-6 years before they die, and are than calcified by the host. There may be 500-3000 cysticerci in one kilogram of infective pork (pig's flesh). The infected pork has brownish spots and is called the *measly pork.*

Infection to primary host . If the bladderworm or cysticercus is eaten with the pork by man, the bladder is digested and the young scolex attaches itself to the wall of intestine by hooks and suckers. A region of proliferation of neck is formed, which develops progiottids of the adult tapeworm. While anchoring itself to the wall of small intestine in matures in two or three months.

Pathogenesis

Both aduts as well as cysticercus stages are harmful :

1. *Adult worms* - The scolex of adult worm remains embedded n the mucous lining of alimentary canal for the attachment. The injury

of musoca may lead to bacterial infection. There may be abdominal discomfort, diarrhoea, frequent hunger pangs, nervousness, insomnia, nausea nad epileptic etc. These symptons are due to the presence of adult worm are collectively known as *Taeniasis.*

2. *Cysticercus larvae* - Sometimes, man may accidentally ingest eggs or proglottids with contaminated food. Serious effects may result from cycticercosis in humans. Although they may be found in almost any organ, cysticerci are commonly located in striated muscles, liver, eye and the central nervous system, where they most often are in the meninges and ventricles of brain (*Marquez-Monter,* 1971). In an aberrant from of *T. solium*, apparently found only in the central nervous system, a cysticercus has daughter cysticerci budding from the bladder wall and forming an invasive, proliferating parasite. This type of metacestode is referred to as a *racemose cysticercus*. No viable scolices have been found in recemose cysticerci (*Beaver* and *Rolon*, 1981).

Living cysticerci usually elicit little inflammatory response. They damage the host by forming space - occupying leisons, that is, their growth puts pressure on the adjacent tissue and interfers with its function or causes necrosis due to reduced blood circulation. In the brain ventricle the organisms cause *hydrocephalus* because the cerebrospinal fluid cannot be properly severe inflammatory reactions.

Treatment

Niclosamide is the drug of choice for intestinal infections, it is highly effective and well tolerated. Carbon tetrachloride and oleoresin are also used as an anthelmintic. For removal of tapeworms, doses of 0.4 gm each of atabrin and acranil or 1 gm of hexylresorcinol with 2 gm of acacia may be given. Cysticercosis is best treated in human by surgical removal of cysts, when possible. If surgery is not possible meberidazol can be administered.

Epidemiology and control

Intestinal infections in humans are acquired by eating inadequately cooked pork. Pigs are infected by eating gravid proglottids passed with human faeces. Since a single proglottid may have hundreds of eggs, a pig may obtain a heavy infection throughout its body by eating only a few proglottids.

Following are the principal means of control :

1. Adequate cooking of food.
2. Freezing pork after slaughter and before it is marketed.
3. Sanitary disposal of human waste.
4. Meat inspection.
5. Prompt treatment of person found to be infected.
6. An educational program aimed at changing habits of personal hygiene so that person do not reinfect themselves or others they have contact with.

PARASITIC ADAPTATION IN *TAENIA*

Taenia is an endoparasite in the intestine of vertebrate hosts. It has undergone a number of structural modifications to adjust in its specific environmnet. The modifications due to parasitic mode of life can be discussed under following headings :

I. Strucural
II. Physiological
III. Modifications in Life-cycle.

I. Structural modificaions can be divided into :

(A) Change in external form
(B) Change in internal form

(A) Change in external form

1. Body is dorso-ventrally flatenned and ribbon like.
2. Neck possess the power of forming new proglottids.
3. Scolex or the anterior end is modified into a hold fast for attachment for this purposes scolex bears suckers and spines.

(B) Change in internal ograns

1. Body wall in devoid of cellular epidermis and remains invested by a thick and resistant cuticle.
2. Alimentary canal is absent because the animal lives in intestine where digested food materials always available.
3. Excretory system consists of protonephridia and which helps in osmoregulation.

4. Respiration anaerobically.
5. Sense organs are absent except tangoreceptors.
6. Highly developed reproductive system.

II. Physiological adaptations

1. It can resist host's digestive enzymes through some unknown mechanism, probably by the secretion of anti-enzymes.
2. It maintains an equilibrium with the osmotic pressure of host.
3. It stimulates the mucous membrane of the intestine of host to secrete a large amount of mucous that cover for further protection against host's digestive enzymes.
4. Power of anaerobic respiration enables it to live in the oxygen-free intestinal contents.
5. Long flatenned body provides large surface area for absorption of digested food of the host that is why the mouth and alimentary canal are absent.

III. Modifications in Life-cycle

1. To cover the risk of finding individuals of different sexes in one host the organism and moreover, each segment is bixexual and self-fertilization occurs.
2. Each proglottid produces 30-40 thousand eggs.
3. A secondary host i.e. pig is introduced to ensure its dispersal from one primary host to another.

ECHINOCOCCUS GRANULOSUS

Echinococcus granulosus or the 'hydatid worm', primarily parasetizes members of the canine family. The larval stage, the hydatid, is normally a parasite of sheep, but almost any mammal will serve as a host. It is potentially dangerous to human beings, in whom it may give rise to a fatal metastasizing, cancer-like tumor.

SYSTEMATIC POSITION

Phylum	-	**Platyhelminthes**
Class	-	**Cestoda**
Sub-class	-	**Eucestoda**
Order	-	**Cyclophyllidea**
Genus	-	***Echinococcus***
Species	-	***granulosus***

History and Habitat

The hydatid disease, which is applied for any tumor or swelling of a cystic feature caused by *Echinococcus*. *Redi* (1684) and *Tyson* (1691) who had reported actual animal nature of hydatid. But it was first discovered by *Goeze* (1782). Next *Hartmann* (1695) who found the adult in the intestine of dog.

Thus *Echinococcus* is generally found in the small intestine of dogs, wolfs, foxes and other carnivorous mammals. But larvae of this parasite develop into hydratid cysts mainly in human being and domestic individuals.

Distribution

Echinococcus is cosmopolitan in distribution, especially in sheep and cattle-raising areas. It has been reported from Iceland, South

Australia, Tasmania, New Zealand, southern South America, southern and northern Africa, southern and eastern Europe, Siberia, Mangolia, northern China, Japan, and the near East. In India, it has been reported from Punjab.

External Features

Shape and Size. The adult of *E. granulosus* is typically cyclophyllidean (*Taenia*), but it differs from it in size. It measures from 2-8 mm. in length.

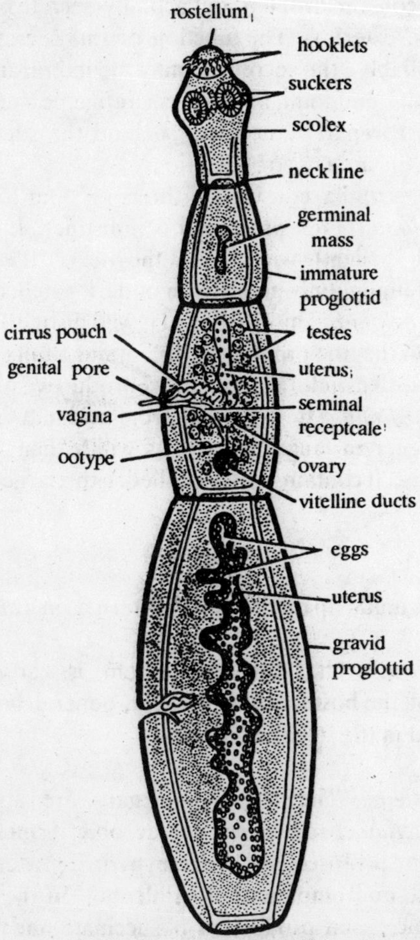

Fig. 7.1. *Echinococcus granulosus*. Entire.

Division of body. The body is divisible into (i) Scolex (ii) Neck (iii) Strobila.

(i) *Scolex.* The scolex is anteriormost part of body measuring 500 u in diameter. The scolex bears four cup-like muscular structures, the *suckers* for the attachment to the intestine of host. Anteriorly situated, rectractile *rostellum* is armed with a double row of 20-80 hooks. *Smyth* (1963) has reported the presence of a group of cells lying just beneath the anterior tip of the rostellum. These cells not only contain droplets but actually secrete small viscid droplets to the exterior. The function of this secretion is still unknown probably the secretion may be important from the immunological standpoint, acting as an antigenic agent.

(ii) *Neck.* The narrow part of the scolex, behind the suckers, forms the neck. It is the area of strobilization.

(iii) *Strobila.* The strobila consists of three or four proglottids.

(a) *First proglottid.* First proglottid is indistinet. It follows the neck and is slightly wider than the neck. It consists of a genital rudiment only in the form of dark patch of cells.

(b) *Second proglottid.* This proglottid is well distinet. It is much larger than the first proglottid. It contains a full compliment of taenioid-like male and female reproductive organs.

(c) *Third proglottid.* It is gravid proglottid and is comparatively bigger in length as well as width than the other proglottids. It contains a uterus filled with a large number of eggs.

Internal Structure.

Body wall. It is made up on the same pattern as in *Taenia.*

Digestive System. The digestive system is absent. The predigested food of the host is taken through general body surface. The food is reserved in the form of glycogen.

Respiratory System. The respiratory organs are absent. The protoscolices of *Echinococcus* carry out aerobic fermentation of carbohydrates. The protoscolices excrete pyruvic, acetic, succinic and lactic acids, and small amount of ethyl alcohol. In the absence of oxygen there is an increased production of succinate and no pyruvate is excreted. Carbon dioxide which is formed as byproduct is diffused out.

Excretory System. It is well developed containing longitudinal and transverse excretory canals with flame cells. Excretory system mainly help in regulating the amount of water in body, hence they are mainly osmo regulatory in function.

Nervous System. Well concentrated in the scolex.

Reproductive System.

Echinococcus is hermaphrodite. The male and female reproductive organs are present in the second proglottid or mature proglottid.

Male Reproductive System. In the middle region of proglottid, encircling the other reproductive organs, are about 20-30 spherical structure known as *testes.* From each testis a fine duct arises known as *vas efferens.* In the postmedial portion of the proglottid the vasa efferentia join to the base of a vas deferens. Vas deferens is an extremely convoluted, sperm filled tube, which passes lateral to an enlarged, coiled, marginal *cirrus* or penis. The cirrus lies in a bulbous *cirrus pouch* or *sac.* The cirrus, during copulation is protruded through a lateral *genital atrium* and genital pore.

Female Reproductive System. Opening into the genital atrium closely behind the cirrus is a female copulatory tube, the *vagina.* The vaginal opening is guarded by a muscular sphincter. Beyond the sphincter, the vagina passes downwards and enlarges at its inner end to form a sperm filled sac, the *seminal recepticle.* On the posterior side of the proglottid there is a bilobed *ovary* with a narrow, central *isthmus* from which an oviduct arises posteriorly. The oviduct joins a duct from the seminal receptical, and this combined fertilization duct continuous posteriorly into a median compact *ootype.* The ootype is surrounded by unicellular *Mehlis' gland.* In the posterior most region there lies a bilobed *vitelline gland.* The two lobes of vitelline glands are oriented dorsoventrally from the vitelline gland a median *vitelline duct* arises that joins the ootype. From the ootype a tube-like *uterus* arises which extends to a point near the anterior border of the proglottid.

Eggs. Eggs, resembling those of taenoid, are monoperculate and measuring about 30x38 u. These are ectolecithal i.e. the zygote is surrounded by the yolk cells provided by vitelline gland and shall material. After an egg-shall is laid down the eggs pass from the ootype into the uterus. When the uterus of a gravid proglottid bursts, the eggs are liberated and passed out along with faeces of dogs. These eggs gain enterence into the intermediate host (man or other mammal) either through water or in forage. Each egg contains a fully developed *onchosphere.* Human infection develops from too intimate an association with dogs. Infection of children is commonly established when dogs are permitted to lick their months. Eggs can be ingested from contaminated fingers.

The eggs, upon being swallowed by a mammalian intermediate host, hatches in duodenum as a typical six-hooked onchosphere. The matching occurs in slight alkaline medium. Activation of embryos take place in either human bile or intestinal juice of ruminants, they are not activated in bile or intestinal juice of dogs and cats. The escaping onchospheres penetrate the intestinal wall, enter the mesenteric venules, and become lodged in the capillary beds of various organs. The hooks are now shed. After three hours of egestion, the embryo reaches its final place of development. Among moose, deer etc. they reach lungs where as in man and other mammals, they reach in the liver, lungs, brain, kidneys, spleen etc. but liver being the most common site. Implantation and growth occur here, Onchosphere now develops into *hydatid cyst.* The cyst formation is a slow process, with the host laying down an envelope of connective tissue around the parasite. In about 4-5 months, the cyst reaches approximately 10 mm. in diameter. It goes on increasing in size and eventually becomes as big as the orange or even sometimes a football.

The various types of larval cysts, formed in the life-history of *Echinococcus* are :

1. *Unilocular hydatid cyst.* The hydatid cyst is composed of three layers. The inner most layer is called *endocyst.* It is thin syncytium forming the germinal layer. The middle layer is called *ectocyst.* It is made up of cuticle which is amorphous and laminated. The outer layer is *pericyst.* It is composed of connective tissue and fibres secreted by the host.

   The cyst produces daughter cysts upon invagination of its wall. These daughter cysts project into the lumen of mother cyst. The cavities are filled with a sterile fluid known as *hydatid fluid.* Within some daughter cysts, tertiary or grand-daughter cysts, form by invagination. These are endogenous or exogenous cysts or brood capsules.

The endogenous cysts fall free into the hydatid fluid of mother cyst white the exogenous cysts, after detaching, migrate to other part of body, where they continue to grow further. The daughter cysts have a fibrous cyst wall internally lined with endocyst.

Rounded and hollow brood capsules are also continuously budded off from the endocyst of mother cyst. These capsules remain attach for some times by means of a narrow pedicle.

From the wall of these brood capsules, minute scolices are formed. Scolex formation may occur from the invagination of walls either of daughter or grand-daughter cyst.

The anterior end is inturned to form a central canal into which the four suckers open and into the expanded base of which the hooklets project. The body of each scolex is filled mesenchymal calls aggregated to form a subcuticular layer and to form a compact cup-shaped mass below the hooklets. The syncytium of the endocyst retains its embryonic potentialities, and new brood capsules may be produced indefinitely. Hydatids thus may expand to enormous size and contains nearly 50 litres of fluid as well as thousands of scoleces. The pressure so produce may severely damage adjacent host organs.

When the infective brood capsule is ingested by the final host, each scolex quickly evaginate and attach to the duodenal mucosa. For the first four to five days in the host they undergo little growth. By the 14th day the scolex has grown in size, and first proglottid has appeared. The genital anlage and a second proglottid appear by the 18th day. Growth continues and fully developed shelled embryos are formed by about the 35th day, and the life cycle is begun again.

2. *Multilocular cyst.* It is characterized by many small bladders embedded in a common enclosing membrane. These are found in cattles. It is considered to be the result of differential reaction inside the host.
3. *Alveolar hydatid in man.* It is characterized by a malignant type of growth with jelly-filled proliferating vesicles embedded in a common dense stroma. The very thin laminated membrane does not restrict the germinal membrane that grows out into surrounding host tissues. It is commonly sterile and it undergoes degeneration and calcification.

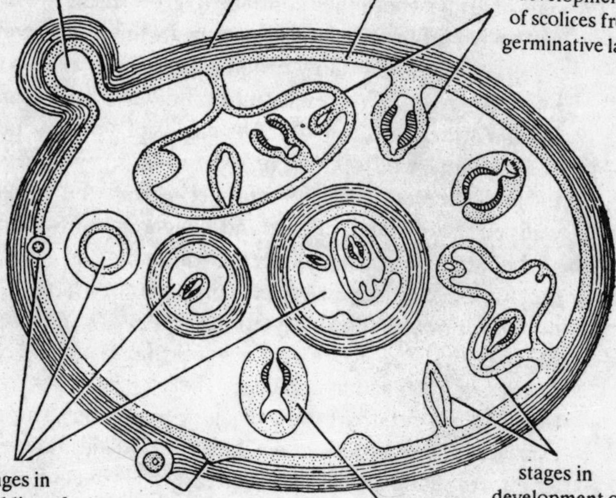

Fig. 7.2. *Echinococcus granulosus*. Hydatid cyst.

4. *Multivesicular.* It is characterized by many adjoining and connected bladders, each having its own separate germinal membrane.

Pathogenesis

Bladders or cysts are harmful when they develop rapidly. It causes inflammatory reaction in the surrounding tissue of the host result in the formation of fibrous tissue. The escaping hydatic fluid not only causing diarrhoea, abdominal pain etc. but also death of the host sometimes.

Treatment

Surgery is the recommended treatment for hydatid disease, but removal of a hydated cyst must be approached with care. An individual harboring a hydatid often develops hypersensitivity to the fluid in the cyst and if the cyst is broken during surgery and fluid leaks into the body cavity, the patient may die within minutes from anaphylatic shock. If the patient survives, protoscolices, which are liberated from breakage of cyst, may set up secondary infection. In order to check the secondary infection sometimes it is recommended to inject the cyst with alcohol or iodine to kill protoscoleces.

Mebendazol has been successfully used in some individuals who have hydatid in locations that are inoperable (*Keystone* and *Murdoch*, 1978). Adult worms in dogs are susceptible to treatment with praziquantel (*Wilson* and *Rausch*, 1980).

Control

The following are the elements of a control programme for hydatid disease :

1. Treatment of infected dogs with an effective anthelmintic or to kill them.
2. Keeping the viscera of domestic animals from being eaten by dogs.
3. Children should not be allowed to play dogs frequently.
4. Avoiding play with dogs and hands should be thoroushly washed before eating.

8

PHYLUM - ASCHELMINTHES

There is variety of opanions among zoologists about the systematic position of nematodes. Some authors (*Gagenbaur* etc.) have designated a separate phylum Nemathelminthes, which includes or not include Acanthocephale and Nematomorpha. Recently, nematodes have been grouped in a class in the phylum Aschelminthes. The term Aschelminthes is coined by *Grobben*.

CHARACTERS

1. Round worms have elongated, cylindrical and vermiform body with tapering ends.
2. Body is unsegmented but may be wrinkled. It is not distinguished into regions.
3. Body unpigmented being either white or with a yellowish tinge.
4. Caudal end of the body generally straight in female but coiled in males and the males are shorter than females.
5. Anterior cephalization is not prominent, therefore, the body is without any definite regions.
6. Mouth terminal surrounded by lips. In strongylodes the lips are modified into teeth known as leaf crown.
7. Amphids and papillae are the main sensory organs and are of great taxonomic value in case in free-living forms.
8. Marine nematodes are annulated due to the presence of transverse striations or longitudinal striations which are of common occurrence.
9. Body is covered by rough resistant cuticle, having bristles, spines warts and papillae etc. Sometimes cuticle is modified as cephalic, cervical and lateral cuticle and in the form of vesicle.
10. Caudal end with a pouch of cuticular nature known as phasmids, very common in parasitic form.

11. Cuticle in the strongyloides is modified into an umbrella like-form known as bursa which is generally supported by muscular rays.
12. The layers of body wall generally made up of cuticle sub-cuticle or epidermis and muscle layer. The cuticle is generally of keratin which allows only glucose and urea etc.
13. Epidermis is syncytial where they are usually divided in four sections in the four longitudinal chords. One dorsal, one ventral and 2 lateral in position.
14. Muscles consist of longitudinal muscle fibres with variable arrangements which is generally holomyarian, meromyarian and polymyarian etc. The muscles are of great taxonomic value.
15. The body cavity is pseudocoel.
16. Digestive tract well developed generally made up of mouth, buccal cavity, pharynx or oesophagus and of various type, intestine and anus.
17. Nervous system consist of a nerve-ring encircling the oesophagus. From it nerves are given out anteriorly and posteriorly.
18. Protonephridia absent but the excretory system is made up of canals or gland-like organs.
19. Sexes are separate. Testes and ovaries are tubular and coiled. Usually there is a single testis. The ducts from the testes open into the cloaca and is associated with accessory structures such as circular spicules. Ovaries, oviduct and uteri are double.
20. Aschelminthes are ovo-viviparous, oviparous or viviparous.
21. Life-cycle complicated may be with or without intermediate host.
22. Larval stages four. Third stage of larva is infective.

Classification

Phylum Aschelminthes is divisible into one class-Nematoda.

Class - Nematoda

1. Body is slender, cylindrical, tapering towards both the ends.
2. Body is covered with cuticle.
3. Intestine is well formed.
4. Body cavity is not lined with epithelium.
5. Cloaca is absent in female.
6. Male and female reproductive organs are well developed.

The Classification of Nematoda is not stable as they have been studied mainly independently. The scheme given below will be found simple and convenient.

Class Nematoda is divided into two sub-classes :

Sub-class I - Aphasmidea

Sub-class II- Phasmidea

Sub-Class I - Aphasmidea

1. Phasmids or caudal sensory organs are absent.
2. Amphids or anterior sensory organs many.
3. Excretory organs either reduced or absent.
4. Coelomocytes and mesenterial tissue are well developed.
5. Candal alae in females are absent, free living.

The Sub-class is divided into following orders :

Order (1) Chromadoridea

1. Oesophagus is not elongated but differentiated into three region.
2. Cuticle smooth or ringed. Ornamented and sometimes with bristles.
3. Amphids spiral.
4. Mainly marine and free-living.
   Examples : *Druconematids, Walichoanolaimus*

Order (2) Enoploidea

1. Mainly marine and free-living.
2. Six sensory papillae on the lip.
3. Ten to twelve sensory bristles arranged in one or two circlets.
4. Cyathiform (goblet-shaped) amphids.
5. A pair of cephalic slits.
6. Size small to moderate.
   Examples : *Pelagonema, Enoplus, Metonchidiamus.*

Order (3) Doyrlaimoridea

1. Found in fresh-water and soil.
2. Cuticle smooth without bristles.
3. Head with an inner circlet of six and an outer circlet of ten sensory papillae.
4. Odontostylet, the piercing apparatus derived from a modified tooth in the oral region.
5. Amphids cyathiform.
6. Posterior part enlarged.
   Examples : *Dorylaimus, Xiphinema.*

Order (4) Mermithoidea

1. Anterior end with 16 labial papillae.
2. Sensory bristle absent.
3. Cuticle is smooth.
4. Amphids either rudimentary or cyathiform.
5. Oesophagus opening in blind intestine.
6. Larvae parasites in invertebrates whereas adults are free living.
   Examples : *Mermis, Paramermis.*

Order (5) Trichiuroidea

1. Intestinal muscles, caudal gland or suckers are absent.
2. Mouth without lips.
3. Pharynx very long and narrow.
4. Females with one ovary.
5. Males with one or no copulatory spicule.
   Examples : *Trichinella, Trichuris.*

Order (6) Desmocolecoidea

1. Small sized with heavily ringed cuticle.
2. The head is armoured. Anterior end with four sensory bristles.
3. Amphids are crescent shaped or pump-shaped.
4. They are marine and free-living.
   Examples : *Tricoma, Desmoscolex.*

Order (7) Monohysteroidea

1. Small forms with smooth or ringed cuticle having bristles.
2. Amphid circular.
3. Anterior end with 4,6,8, or many sensory bristles.
4. Free living forms, few are marine, some are fresh water and others terrestrial.
   Examples : *Monohystera, Siphonolaimus.*

Sub-class II - Phasmida

1. With phasmids and pore like amphids.
2. Excretory organs with paired lateral canals.
3. Coelomocytes 4-6.
4. Caudal or hypodermal glands absent.
5. Reproductive system well developed.

The sub-class has been divided into following orders :

Order (1) Rhabditoidea

1. Mouth without lips and with 6,12, or 18 papillae.
2. Cuticle is ringed or smooth.
3. Pharynx, long with posterior bulb.
4. Single ovary in females and males with copulatory spicules.
5. Mostly parasite on birds and mammals.
   Examples : *Rhabditis, Diplogaster.*

Order (2) Oxyuroidea

1. Pinshaped small or medium sized.
2. Cervical papillae absent.
3. Pharynx with posterior bulb.
4. Females with slender tail.
5. Males with one or two copulatory vesicles and sometimes with alae forming a bursa.
6. Parasites in invertebrates and vertebrates.
   Examples : *Enterobius, Oxyuris.*

Order (3) Ascaroidea

1. Large parasitic.
2. Mouth surrounded by three rarely six lips.
3. Cervical papillae present.
4. Pharynx with or without valve.
5. Female short tail.
6. Male with ventrally coiled tail and with two equal copulatory spicules and usually without alae.
   Example : *Ascaris.*

Order (4) Strongyloidea

1. Mouth without distinct lips but has leaf crowns.
2. Pharynx is without bulb.
3. Males are provided with a distinct caudal bursa which is supported with rays.
4. Females with a ovejactor.
   Examples : *Ancylostoma, Necator, Strongylus.*

Order (5) Dracunculoidea

1. There are no definite lips or buccal capsule.
2. Mouth is surrounded by ring of papillae.
3. Pharyngeal bulb is absent. Pharynx is muscular anteriorly but glandular posteriorly.
4. Copulatory bursa is absent in males.
5. They are parasite of vertebrates.
   Examples : *Dracunculus, Micsopleura.*

Order (6) Spiruroidea

1. Mouth with two lateral lips, sometimes 4 or 6.
2. Pharynx with bulb with anterior muscular and posterior glandular part.
3. Males without bursa but have spirally coiled tail.
4. Parasitic forms.
   Examples : *Gnathostoma, Spiroxys.*

Order (7) Rhabdiasoidea

1. Cuticle smooth.
2. Definite pharyngeal bulb is absent.
3. Parasitic stage is hermaphrodite or parthenogenetic free-living stage may develop into males and females.
   Examples : *Rhabdias.*

Order (8) Filariodea

1. Slender filiform worms.
2. Mouth simple, lips are absent.
3. Females larger having an anterior vulva and a long tubular vagina.
4. Males with two unequal spicules. Alae present or absent.
5. Found in blood or lymphatic system. Transmitted by blood sucking insects.
   Examples : *Wuchereria, Microfilaria, Loa.*

TRICHURIS

Phylum	-	**Aschelminthes**
Class	-	**Nematoda**
Sub-class	-	**Aphasmida**
Order	-	**Trichuroidea**
Genus	-	***Trichuris***

1. It is commonly celled as *whip-worm.* It is endoparasite found in the caecum and large intestine of man.
2. It has world wide distribution but common in tropical countries.
3. Female is slightly larger than males.
4. The anterior part of the body having pharynx is usually more slender and this portion is buried in the mucosa of host's alimentary canal.
5. Mouth is simple without lips.
6. It feeds on blood and epithelial cells.
7. Life-cycle is simple with no intermediate host.
8. Development of eggs take place in moist soil. Embryos develop in about three weeks.

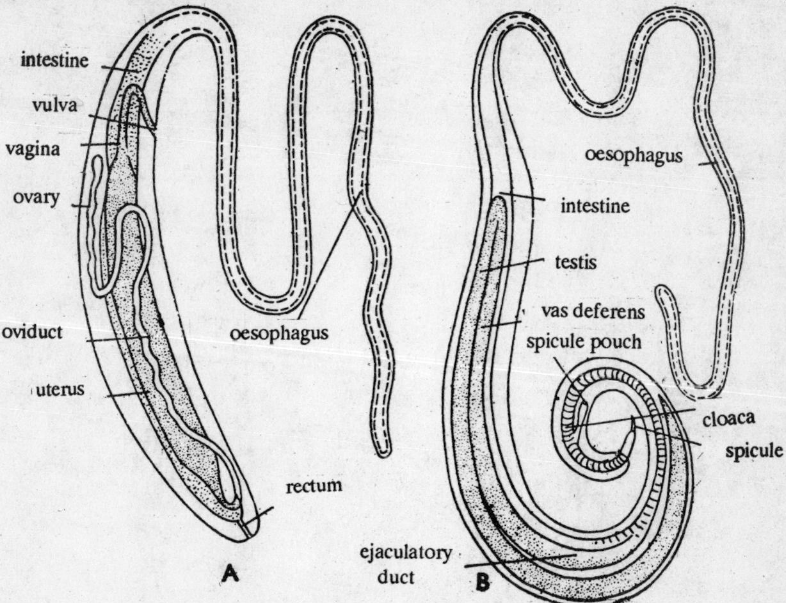

Fig. 8.1. *Trichuris*

9. Contaminated food and water responsible for infection in man by swallowing eggs with juveniles.
10. Liberated juvenile grow into adults within a month.
11. Excretory system is absent.
12. Heavy infections is responsible for diarrhoea and anaemia.

TRICHINELLA

The systematic position is the same as that of *Trichuris*.

1. It is a parasite of man, pig, rat and other vertebrates. It is commonly known as *trichinia worm.*
2. It is commonly found in Europe, United States and Arctic regions.
3. Sexes are separate, males are about 1.5 mm. in length and the females are 3 mm. in length.
4. In both the sexes, the mouth is without lips.
5. Pharynx is long, slender and non-muscular and is embedded in stichosome.

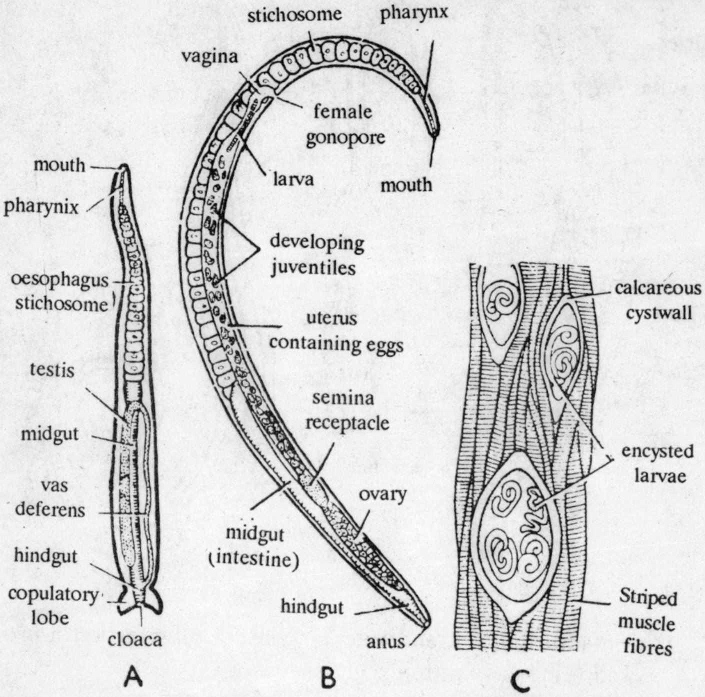

Fig. 8.2. *Trichinella*. A-Adult male. B-Ault female, C-Encysted larvae in striped (voluntary) muscle of pig.

6. The oesophagus is tubular, well developed and occupies 1/3 of the total length of body.
7. The female worms which are viviparous enter the lymphatic vessel and deposit their young ones. Each young worm then reaches the muscular system where they secrete a cyst around them. This is called encysted sexless stage.
8. The encyated sexless trichinae are ingested by man along with pork-flesh, where the cyst wall is dissolved by the action of digestive juice and the reproductive organs start developing.
9. The developing worm begins to migrate into the muscles and produces the symptoms of a dreaded disease called *Trichinosis*.
10. Patients feel gastro-intestinal pain, and weakness with or without fever.

RHABDITIS

Phylum	**-**	**Aschelminthes**
Class	**-**	**Nematoda**
Sub-class	**-**	**Phasmida**
Order	**-**	**Rhabditoidea**
Genus	**-**	***Rhabditis***

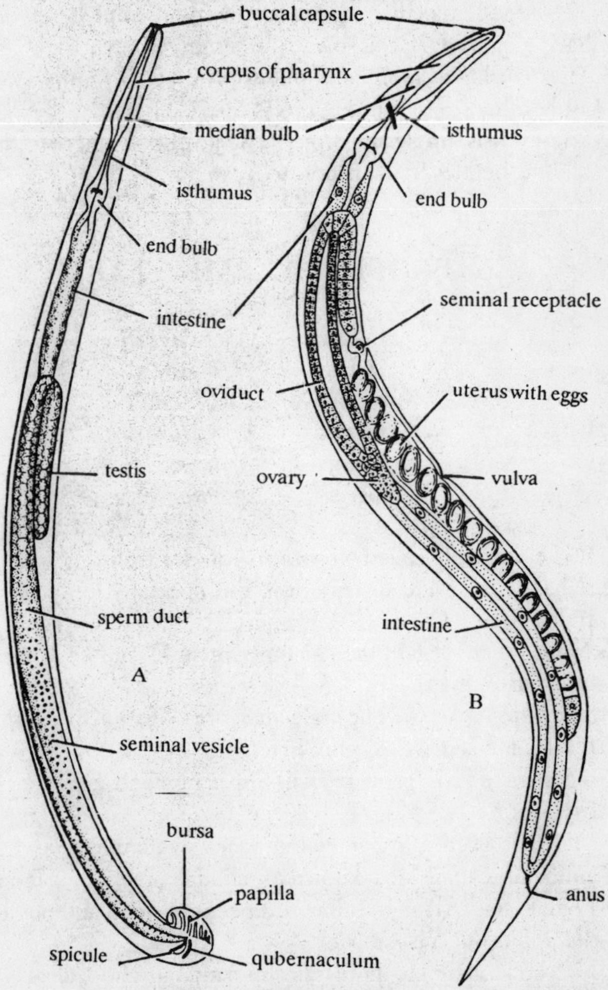

Fig. 8.3. *Rhabditis*
A = male, B = female

1. Genus *Rhabiditis* includes numerous free-living, semi-parasitic forms. They are commonly found in organic matter, faeces of man. They feed on the fluid formed by bacterial decying of organic matters.
2. The pharynx has two bulbs with an isthmus joining them.
3. The excretory system is H-shaped.
4. The reproductive system is J-shaped.
5. The large female has a pointed tail and didelphic sex-organs i.e. having paired ovaries, oviducts, seminal recepticles and uteri.
6. The male has a flatenned copulatory bursa, two equal spicules and monorchic sex organs i.e. single testis.
7. Life-cycle is simple and direct. The juveniles adhere to the bodies of dung beetles for transport to fresh dung.

ENTEROBIUS VERMICULARIS

Phylum - Aschelminthes
Class - Nematoda
Sub-class - Phasmida
Order - Oxyuroidea
Genus - *Enterobius*
Species - *vermicularis*

1. It is commonly called *'pinworm'* or *'seatworm'*.
2. It has world-wide distribution, but specially found in Europe, India, Africa, Canada, Chile and U.S.A.
3. Adult worms live in the calcum, appendix and adjacent parts of intestine of man.
4. Sexes are separate. The male measures 2-5 mm. in length and the female measures 8-13 mm. in length.
5. Anterior end is provided with three lips and a pair of cephalic expansion.
6. Male has a ventrally curved tail with a single spicule and a pair of cuticular expansions known as caudal alae. It is monorchic.
7. Female has a straight tail which becomes long and pointed. It is didelphic Vulva is anterior.
8. Life-cycle is simple there is no intermediate host Gravid females migrate to anus causing itching sensation. They crawl and deposit eggs on the perineal skin. Due to scratching, gravid females rupture and thus small eggs are liberated. The reinfection in

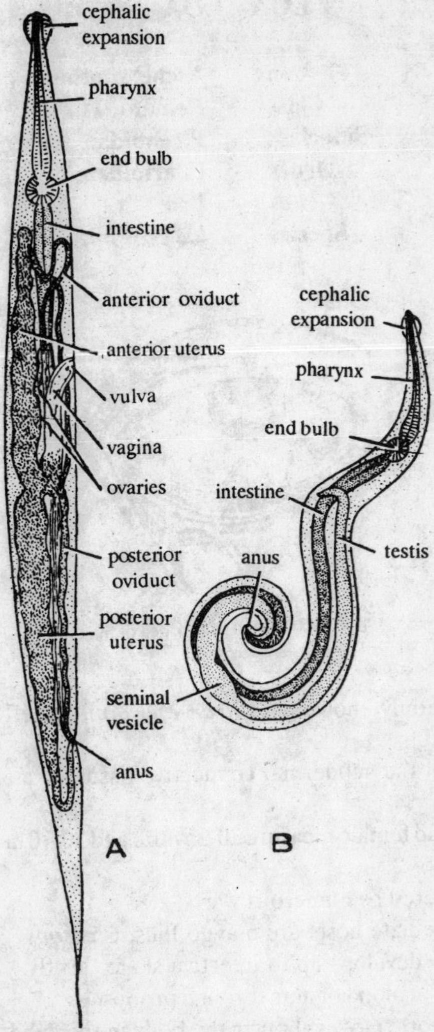

Fig. 8.4. Entrobius
A = male, B = female

children takes by putting fingers in mouth. Eggs hatch. Juveniles inhabit small intestine. They moult twice and turns into adults which reach their normal habitat. The infection is called *Enterobiasis*. Symptoms are inflamation of caecal mucosa, abdominal pain, insomnia, hysteria, restlessness, loss of appetite and sometimes appendicitis.

LOA LOA

Phylum - Aschelminthes
Class - Nematoda
Sub-class - Phasmida
Order - Filarioidea
Genus - *Loa*
Species - *loa*

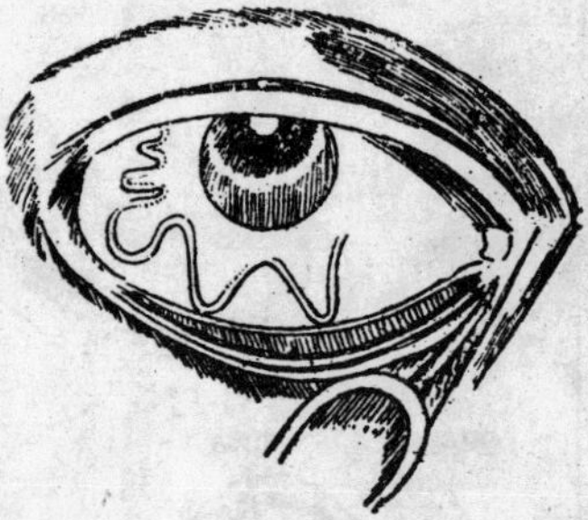

Fig. 8.5. Loa loa in the cornea of man.

1. It is commonly known as *African eyeworm,* chiefly found in Africa.
2. They live in the subdermal connective tissue of man in West Africa.
3. The male and female measure 20-35 mm. and 20-70 mm. in length respectively.
4. Body is covered by numerous warts.
5. The intermediate hosts are mango flies, *Chrysops.* The larvae moult and develop up to infective stage in fly. These are transmitted in blood of man through proboscis.
6. The adult worm moves about in the body in the subdermal tissue and sometimes passes across the eye ball causing intense itching and swelling known as 'calabar swellings'.
7. These are injurious and fatal when they penetrate brain and spinal cord.

DRACUNCULUS

Phylum	-	**Aschelminthes**
Class	-	**Nematoda**
Sub-class	-	**Phasmida**
Order	-	**Dracunculoidea**
Genus	-	***Dracunculus***
Species	-	***medinesis***

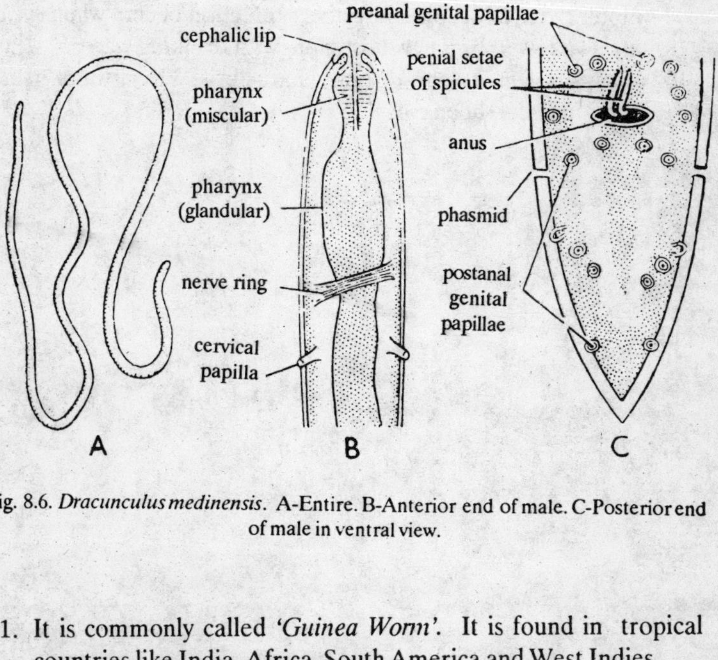

Fig. 8.6. *Dracunculus medinensis*. A-Entire. B-Anterior end of male. C-Posterior end of male in ventral view.

1. It is commonly called *'Guinea Worm'*. It is found in tropical countries like India, Africa, South America and West Indies.
2. It lives in the deeper subcutaneous tissues of man.
3. The anterior end is blunt white the posterior end is pointed.
4. The adult male measures 4 cm. in length. It's hind end is curved and bears 10 pairs of genital papillae and a pair of pineal spicules.
5. Adult female measures 70-120 cm. in length. It's hind end is straight and bears a spine-like process.
6. The oesophagus is distinguished into an anterior narrow muscular region and a posterior broader region glandular region.
7. The female reproductive system is didelphic with opposed uteri.

8. Females are viviparous. The alimentary canal and vulva are atrophied in mature females.
9. Mature female migrates to the superficial layers of the skin in those parts of the body which are likely to come in contact with cold water. Toxic substances are produced under the skin which results in the formation of a blister, leading to an ulcer, which breaks on coming in contact with water. On coming in contact with water, the worm ruptures, releasing larvae in water. The larvae swims about but do not develop further unless swallowed by *Cyclops,* which serves as intermediate host. Here the larvae moult twice & become infective. Infection occurs when *cyclops* are swallowed by a new host alongwith drinking water.
10. It causes *guinea worm disease* characterised by asthma, nausea, vomiting, diarrhoea etc.

9

ASCARIS LUMBRICOIDES

Much is known about the morphology and life-cycle of Nematoda but out of them *Ascaris* is most known. *A. lumbricoides* is the most common nematode parasite found in the human intestine. Being of large size and found abundently, *Ascaris* is studied as the main example of the nematodes.

SYSTEMATIC POSITION

Phylum	-	**Aschelminthes**
Class	-	**Nematoda**
Sub-class	-	**Phasmida**
Order	-	**Ascaroidea**
Genus	-	***Ascaris***
Species	-	***lumbricoides***

Habit and Habitat

A. lumbricoides, a large nematode parasite of man, is world wide in distribution. It occurs more commonly in the small intestine of children than in adults because of the children's custom of playing in contaminated, egg-laden soil adjacent to dwellings. As many as 1000 - 1500 adult worms may be present in a single host. Because of their strong, pointed and cuticulate bodies and their habit of wandering, they may perforate the gut wall and cause peritonitis or they may enter the bile or pancreatic ducts or the vermiform appendix and interfere with normal digestive functions.

External Features

Shape and Size. Both the sexes are elongated, cylindrical and pointed at both the ends. Males 15-31 cm. long and 2-4 mm. in diameters, while the females are 20-35 cm. long and 4-6 mm. in width.

The males are further distinguished by a ventrally curved posterior end, from the slit-like and opening of which often protrude two copulatory spicules.

Colouration. In the living condition, the colour of the animal is milky-white or reddish-yellow. There is a characteristic lustre on the body.

Longitudinal lines. There are four longitudinal lines or streaks, running throughout the length. One is mid-dorsal, one mid-ventral and two lateral. The dorsal and ventral lines are white in colour, the later streaks are thicker and broader than dorsal and ventral streaks and are brown in colour.

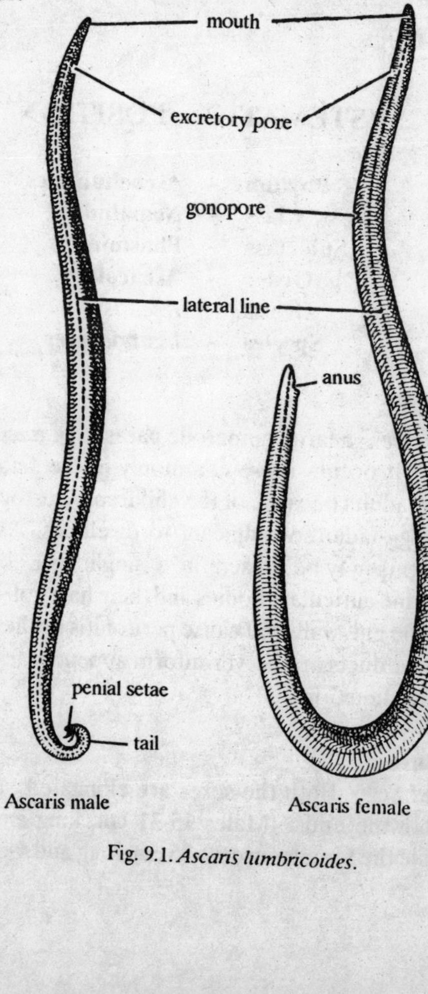

Fig. 9.1. *Ascaris lumbricoides.*

Mouth. A triangular mouth is present at the anterior end of the body. It is surrounded by three denticulated lips. One of these is mid-dorsal and other two are latero-ventral in position. The mid-dorsal lip is almost elliptical and has a forked fleshy core in the centre. A circle of four *sensory papillae* is found on it. The remaining two lips also have a forked core and two rows of sensory papillae. Each sensory structure is called *amplids.* The sensory papillae of nematodes are very useful in the identification of the species and assignting their position in the classification. The sensory papillae are *tangoreceptors.* The amphids are *chaemoreceptors.*

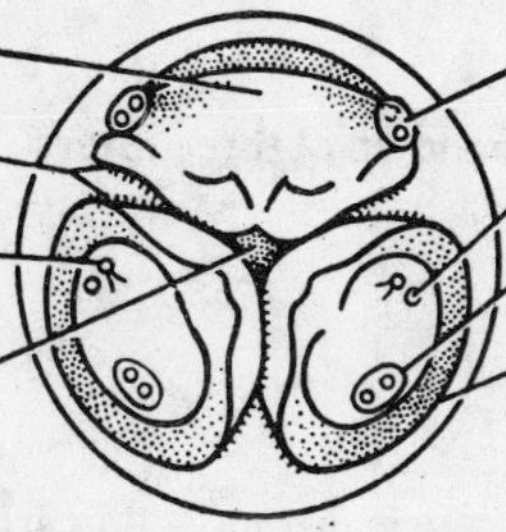

Fig. 9.2. *Ascaris.* En-face view of lips and mouth.

The posterior end of the male bears ventrally, 50 pairs of *pre-anal papillae* in front of the cloacal aperture and five pairs of *post-anal papillae* behind it. The first two pairs of post-anal papillae are double. These papillae of the male help in copulation.

Excretory pore. It is situated mid-ventrally, a little behind the mouth.

Genital pore. It is present only in female and is situated mid-ventrally at about 1/3 distance from the anterior end.

Anus. It is also present only in female. It is a transverse pore quarded by thick lips, lies mid-ventrally a little infront of posterior end.

Cloacal aperture. It is present only in male. It is situated mid-ventrally in front of the curved posterior end. Two chitinous *penial setae* or *penial spicules* are protruding through cloacal aperture.

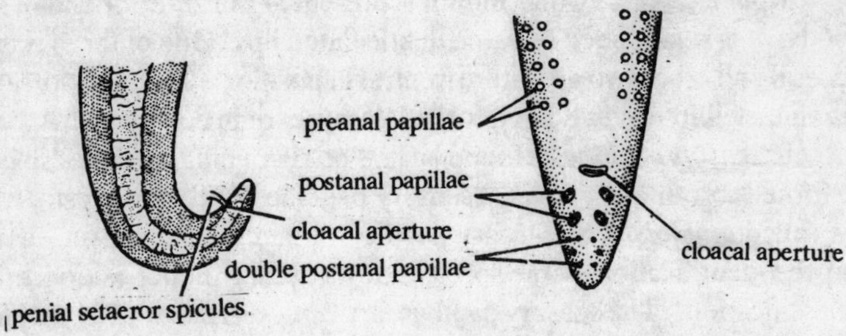

Fig. 9.3. *Ascaris*. Posterior end in lateral view and ventral view.

Body Wall

The body wall is made up of three layers

(1) Cuticle
(2) Epidermis
(3) Muscle layer

1. *Cuticle.* The cuticle forms a thin transparent, elastic but resistant covering around the body. It protects the body against the action of digestive juices is secreted by underlying epidermis. Histologically, the cuticle is formed of several layers and can be distinguished into (i) Cortical layer, (ii) Matrix, (iii) Fibrous layer and (iv) Basement membrane.
   (i) *Cortical layer.* It is made up of dense keratin. It protects the animal from the digestive juices of host.
   (ii) *Matrix.* It is an elastic layer containing sulphur rich matricin.
   (iii) *Fibrous layer.* It is made up of collagen fibres that cross each other obliquely.
   (iv) *Basement membrane.* It forms the inner limit of cuticle.

Under electron-microscope the cuticle is made up of following layers :

(1) *Lipoid layer.* It is in the form of a thin osmophilic membrane, about 1000 A° thick.
(2) *Cortical or cortex layer.* It consists of (a) an *outer cortical layer* which lies as discontinous strips or rings around the body, and (b) an *inner cortical layer.*

(3) *Matrix layer.* It consists of (a) an outer *fibrillar layer* traversed by branching canals, (b) a *homogeneous layer* which several radial striations, and (c) a *boundary layer* resembling the outer cortical layer.

(4) *Fibre layer.* It consists of collagen fibres arranged in three strata.

(5) *Basement membrane.* It is a thin layer surrounding the epidermis.

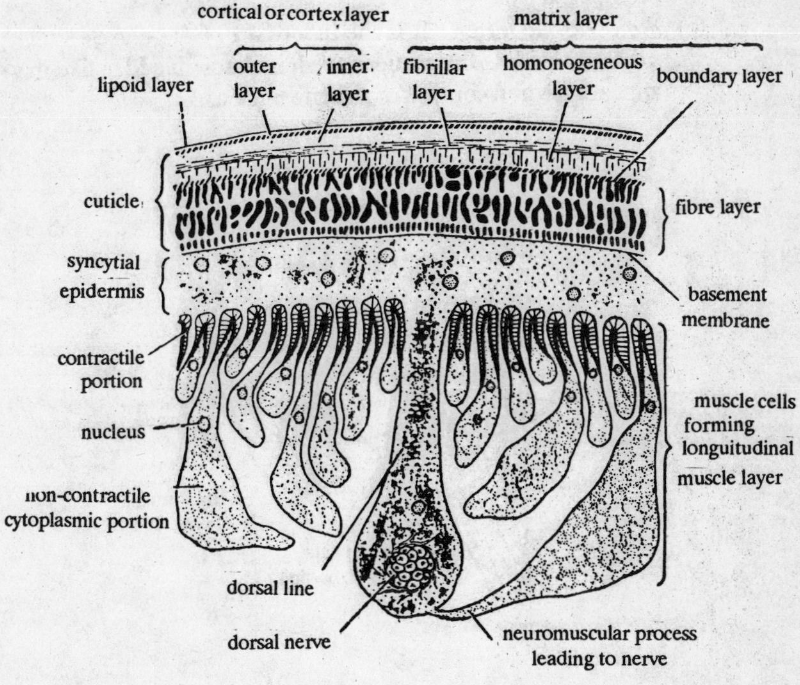

Fig. 9.4. Body wall in section.

1. *Epidermis.* It is *syncytial* in nature as it is having no cellular boundaries i.e. the nuclei are scattered in cytoplasm. It is thickened and projects into the pseudocoel as four longitudinal ridges called, *epidermal chords.* These are dorsal, ventral and lateral in position. The lateral chords are thicker than others, containing longitudinal excretory canals and lateral nerves in them. The dorsal and ventral are thin containing nerve cords and terminate towards the posterior end of the body.

3. *Muscle layer.* Beneath the ectoderm is a single layer of longitudinal muscles which is distinguished into four quadrants. Of them two are dorso-lateral and two ventro-lateral. Each containing about 150 muscle cells. The muscle cells are highly specialised spindle-shaped cells, distinguished into two parts lying below the ectoderm.
   (a) *Fibrillar region.* Formed of contractile and non-contractile fibres arranged alternatively, it is found wrapped around the protoplasmic part.
   (b) *Protoplasmic region.* It is the inner part of the muscle cell situated towards the pseudocoel. It is almost bladder-like in appearance and consists of a single nucleus.

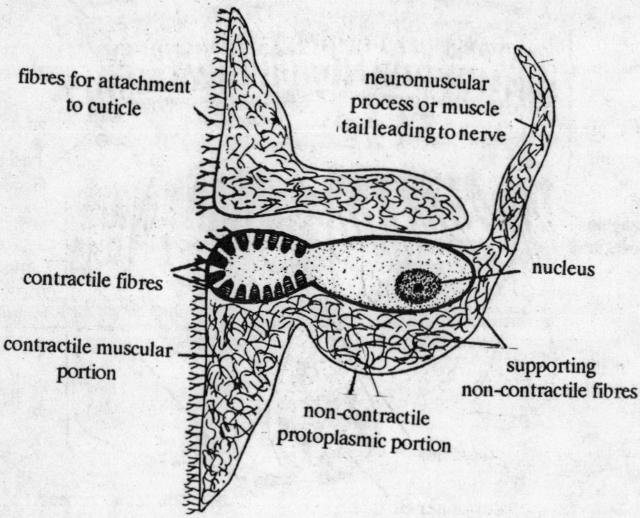

Fig. 9.5. *Ascaris.* A muscle cell in T.S.

There is a complete absence of circular muscles in nematodes.

Pseudocoel.

There is a large fluid filled space between the body wall and the viscera. The fluid contains fibrous tissue in the form of fenestrated membranes and fixed cells. The epithelial lining of this cavity is absent hence cannot be a true coelom. The fenestrated membranes extend over the viscera and muscle layer from a single giant *pseudocoelocyte* placed on the dorsal side where pharynx connects with the intensine. Four more star shaped giant cells lie in relation to epidermal cords. Thus in all there are five pseudocoelocytes. The pseudocoelocytes and fenestrated membranes may be regarded as a kind of mesenchyme.

In fact, the spaces are said to be large vacuoles in a few, enormous cells. A protein-rich fluid of characteristic, unpleasent odor fills the cavity. This fluid may produce some of the toxic and inflammatory symptoms of ascarid infections.

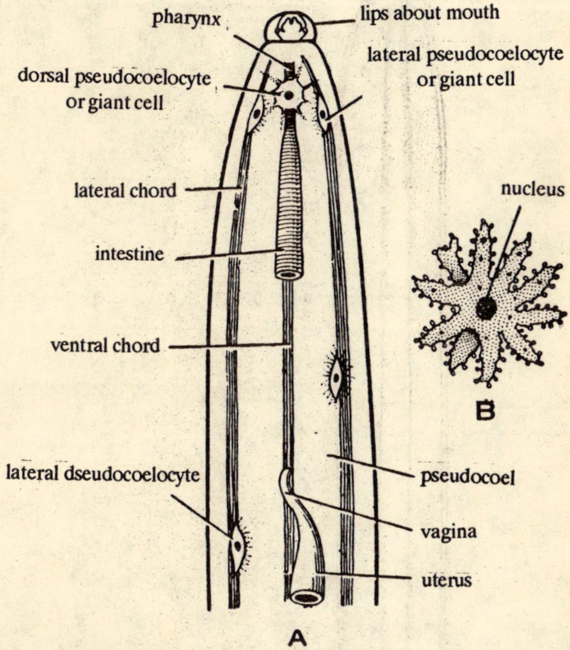

Fig. 9.6. *Ascaris*. A-portion of giant cells in pseudocoel. B-single giant cell.

Locomotion

Locomotion is by means of longitudinal muscles which produce undulating movements of the body in the dorsoventral waves. The circular muscles are absent to counteract against longitudinal muscles, the hydrostatic pressure of body fluid against the cuticle serves the purpose.

Digestive System

The alimentary canal of *Ascaris* is complete. It comprises mouth, pharynx, intestine, rectum and anus.

Mouth. It lies at the anterior end of the body. It is bounded by three lips-one mid dorsal and two latero-ventral in position. It leads into the pharynx.

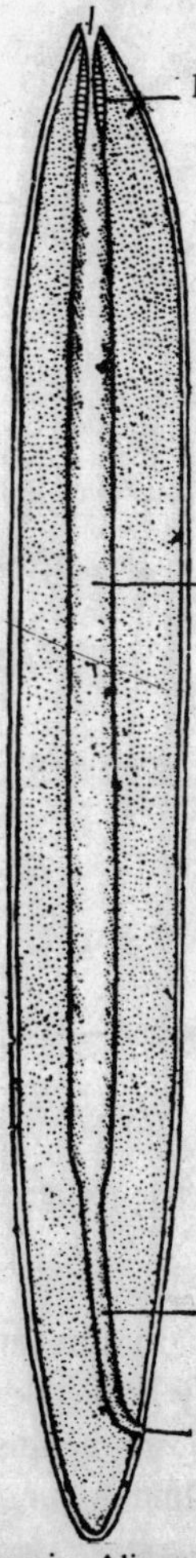

Fig. 9.7. *Ascaris*. Alimentary canal.

Pharynx. The pharynx is muscular, about 10 mm. long. It provided with a triradiate, cuticulate lumen, and has a wall composed of radially arranged, lumen-dilating muscles. A single, pinnately branched dorsal, and a pair of palmately branched subventral, uninucleate glands are embedded in the wall of the pharyns. These are unicellular structures discharging their contents into phanyngeal cavity. The wall is composed of syncytial epithelium. The pharynx is followed by the intestine. The pharynycal wall projects into the intestine as a phanyn-geo-intenstinal valve.

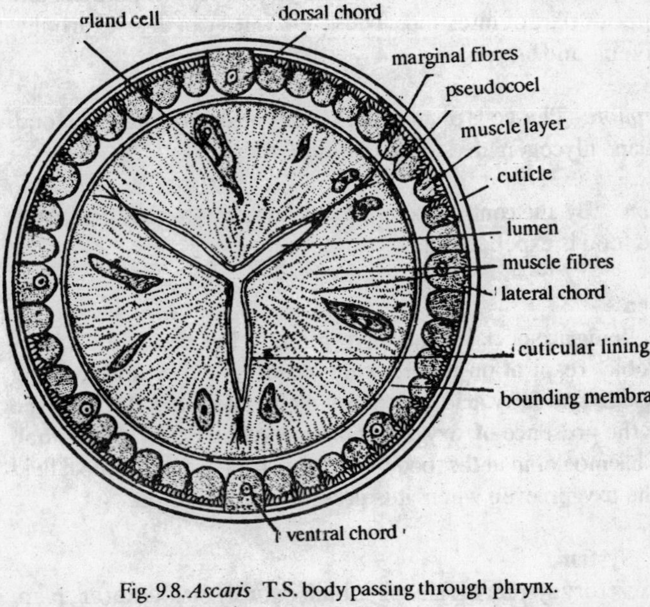

Fig. 9.8. *Ascaris* T.S. body passing through phrynx.

Intestine. It is a long, thin, walled, dorsoventrally flattened tube which is devoid of muscles. The intestinal wall is formed of a single layer of elongated columnar cells lined externally by a *basement membrane* and a thin layer of *cuticle.* The free inner margins of columnar cells are produced into several finger-like projections called the *microvilli.* These form a *brush border* and increase the surface area for absorption.

Rectum. The intestine is followed by a short *rectum* which is also dorso-ventrally flattened. Wall of the rectum consists of tall columnar cells and is lined internally by cuticle and externally by muscle tissue. In male, the rectum opens into the *cloaca* which also receives the ejaculatory duct, while in female, it opens out through the anus.

Anus. The anus is a transverse slit guarded by anterior and posterior lips and is provided with a special muscle, the *depressor ani.*

Digestion. Ascaris gets semi-digested or digested food from its host. So it does not require to digest food. However, it can digest semi-digest food by means of enzymes secreted by gland cells of pharynx and cells of anterior part of intestine. These enzymes can digest carbohydrates, proteins and fats.

Absorption. The cells of intestine generally absorb digested food. Excess fat and glycogen are stored in different tissue.

Egestion. By the contraction of dilator muscles attached to rectum undigested food is expelled out.

Respiration

There being no special organs of respiration, but the parasite carry on anaerobic respiration by breaking down glucose and forming carbon dioxide and fatty acid which are eleminated out through the cuticle. In the presence of oxygen, it respires aerobically. The small amount of haemoglobin in the body wall (muscles) and pseudocoel fluid takes up the oxygen even when it is present in low tension.

Excretory System.

The excretory system of *Ascaris* is built entirely on a different plan. The flame cells or nephridia are absent. Renette cells of osmoregulatory function found in other nematodes are not found here.

In *Ascaris,* the excretory system is made up of a syncytium containing four or more nuclei. It consists of :

1. *Excretory duct.* There is an excretory duct having two or more nuclei embedded into its walls. The excretory duct opens out through excretory pore.
2. *Excretory cell.* There is an excretory cell comprising two lateral excretory filaments running along each side of the body adjacent to the lateral chord.

Fig. 9.9 *Ascaris*. Exretory system.

3. *Commissure.* An excretory bridge is formed where two lateral filaments join. Each excretory filament contains an

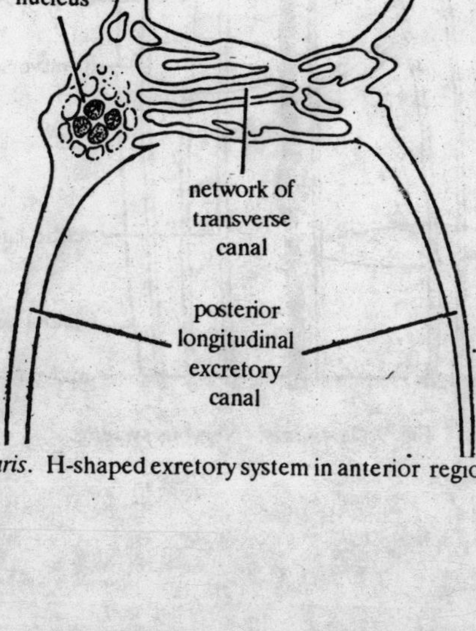

Fig. 9.10. *Ascaris*. H-shaped exretory system in anterior region of body.

intracellular excretory canal. The intracellular excretory canals divide up into a network of canaliculi in the commissural region. Excretory products are believed to diffuse into the canals which secrete urea, which is eleminated out through the excretory pore.

Nervous System

The nervous system consists of ganglia and nerves, all in close association with the ectoderm. The nervous system of *Ascaris* has been thoroughly studied by *Goldschmidt.* The number of cells taking part in the formation of entire system is apparently fixed and it is 162. There is a distinct *circumpharyngeal nerve ring* around the pharynx which connects a pair of *lateral ganglia,* a *dorsal ganglion,* a pair of *subdorsal ganglia* (each sub-divided into six smaller ganglia) and a pair of large *ventral ganglia* from these ganglia of the nerve-ring there

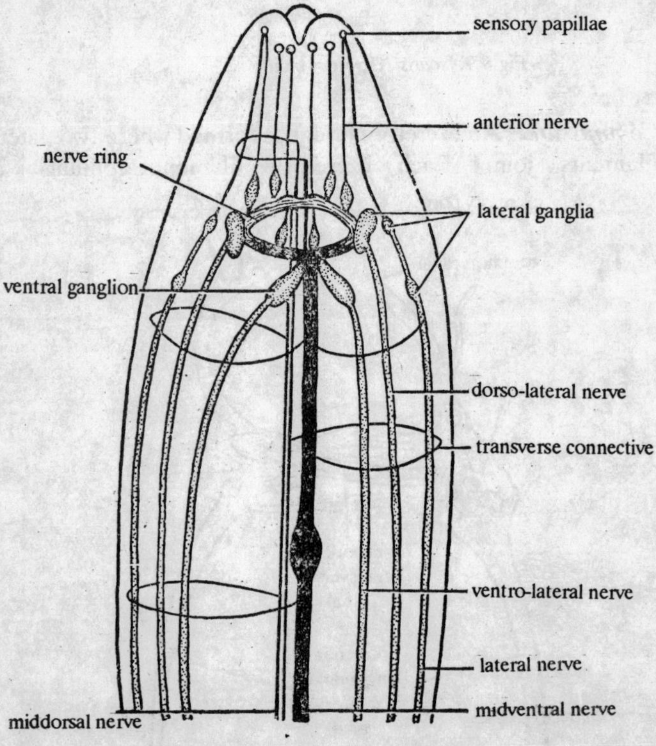

Fig. 9.11. *Ascaris.* Nervous system.

proceed backwards eight nerves : a *mid-dorsal nerve* that passes through the dorsal epidermal chord; a *mid-ventral nerve* that passes through the ventral epidermal chord ; a pair of *dorso-lateral nerves* that run just above the lateral chords ; a pair of *lateral nerves* that extend through the lateral epidermis chords and a pair of *ventro-lateral nerves* that run just below the lateral epidermal chords. Of these, the ventral nerve is most prominant and is ganglionated through out its length, therefore, it is regarded as nerve cord. Posteriorly the nerve cord joins a *pre-anal ganglion.* The dorsal and ventral nerves are connected at intervals by asymmetrical *dorso-ventral connectives* throughout their length. Similarly, the ventral and lateral nerves are also conncctcd at interval by *ventro-lateral commissures.*

One of the six ganglia formed by sub-division of each lateral ganglion is known as *amphidial ganglion* which supply nerves to amphids of its side. The genital papillae of the male worm are innervated from special *bursal nerves.*

Sense organs

The sense organs include papillae amphids and phasmids.

Papillae. Various types of papillae such as *labial, cervical* and *genital papillae* are present in *Ascaris.* A labial papilla consists of a nerve axis placed inside a cuticular elevation surrounded by supporting cells. The dorsal lip has two double papillae, each consisting of a labial papilla on the medial side and a cephalic papilla on outerside. Each

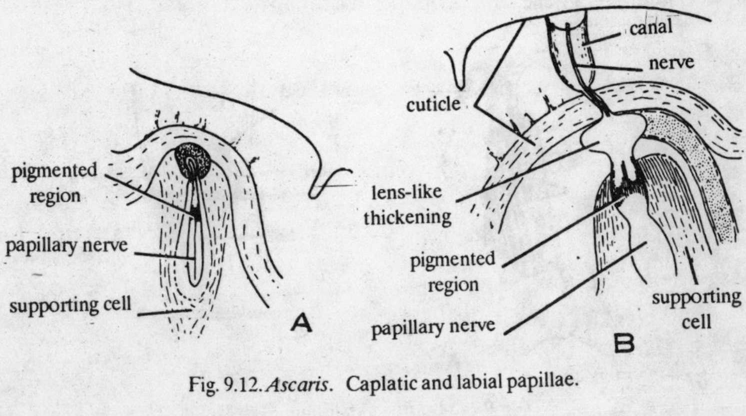

Fig. 9.12. *Ascaris.* Caplatic and labial papillae.

ventro-lateral lip has a labial papilla near its upper margin and a double papilla near its lower margin. The double papilla a labial papilla towards the medial side and a cephalic papilla towards the outer side. Thus the lips bear six labial and four cephalic papillae in all. Genital and caudal papillae have a similar structure.

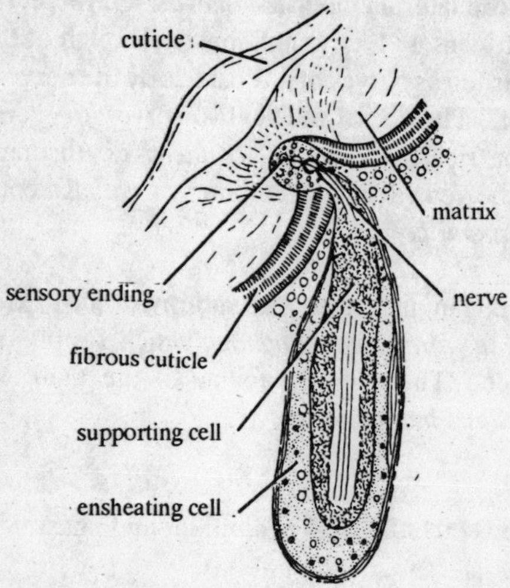

Fig. 9.13. *Ascaris*. Cervical papilla.

Amphids are two in number, one on each ventro-lateral lip a little above the labial papilla. An amphid consists of supporting cell opening out side by a pore and having a bundle of nerve-fibres that unit into one fibre. These are probably a chaemoreceptor.

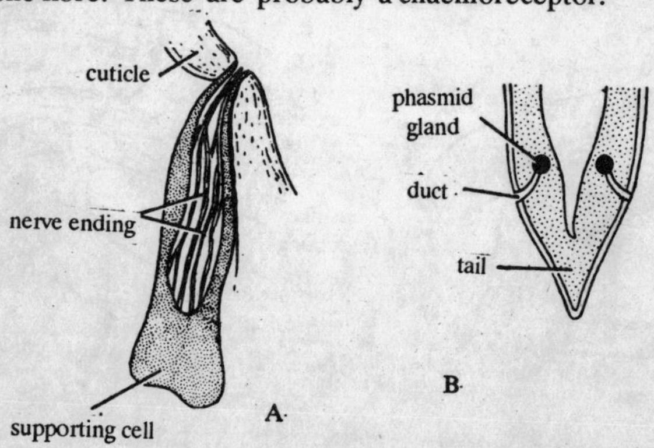

Fig. 9.14. *Ascaris*. A-amphid. B-phasmid.

A pair of *phasmids* occur on each side of the tail. Phasmids is the out let of a unicellular gland opening over a small papilla which is probably chaemoreceptor.

Reproductive System

Sexes are separate and there is a well distinct sexual dimorphism.

I. Male Reproductive System.

The male reproductive organs are :

1. Testis
2. Vas deferens
3. Seminal vesicle

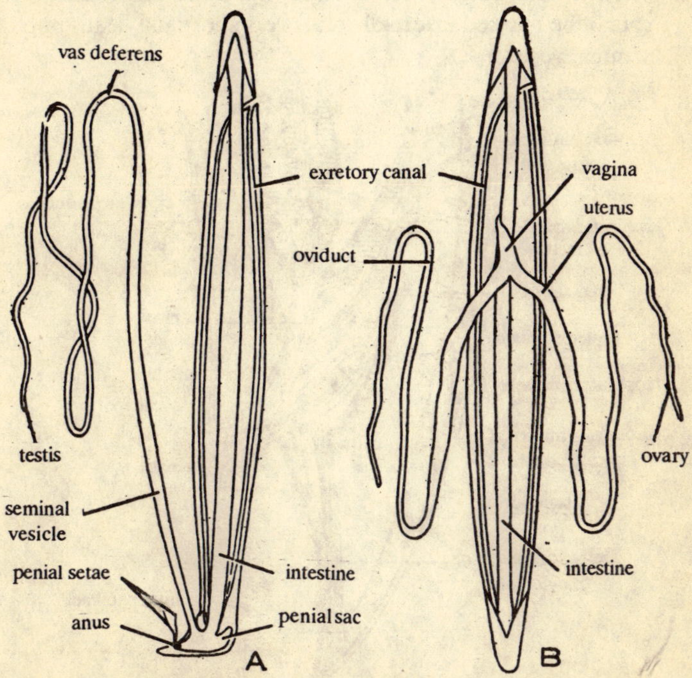

Fig. 9.15. *Ascaris*. A-male reproductive system. B-female reproductive system.

4. Ejaculatory duct
5. Penial sac with penial setae.

1. *Testis.* The testis is a long thin, coiled and thread-like structure in the anterior part of pseudocoel. It represents *monorchic condition* (presence of single testis) and *telogonic* mode of origin of germ cells (*i.e.* germ cells developing only from the proximal part of gonad). The testis can be distinguished into three parts :-

   (a) anterior solid part consisting of sex cells, (b) middle part and (c) Posterior part. The central axis of the tube is in the form of a solid *cytoplasmic rachis,* around which the clusters of amoeboid sperms in various stages of development.

   It is continued posteriorly into the vas deferens.
2. *Vas deferens.* The vas deferens is as wide as the testis and the two cannot be marked externally. It is very short and leads into the seminal vesicle.

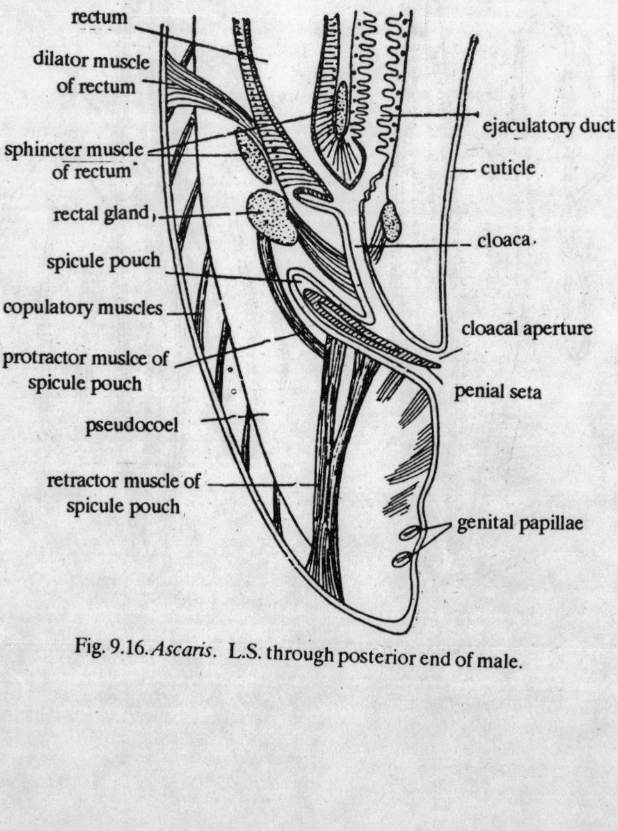

Fig. 9.16. *Ascaris*. L.S. through posterior end of male.

3. *Vesicula seminalis.* The vesicula seminalis is a wide tube present in the posterior 1/3rd part of the body. It joins the ejaculatory duct.
4. *Ejaculatory duct.* It is a short, narrow but muscular tube opening into the cloaca, which leads outside by cloacal opening.
5. *Penial sac with penial setae.* On the dorsal side of the cloaca are a pair of sac-like structures called penial sac. These are formed by evaginations of the cloaca. Each pouch secretes and lodges a club-shaped *spicule* or *penial seta.* The spicule is enclosed in a sheath and consists of a cytoplasmic core surrounded by thick cuitcle. Each spicule measures 2-3.5 mm. in length. The penial sac have special *protractor* and *retractor* muscles from their walls, the muscles push the spicule out and withdraw them back into the pocket.

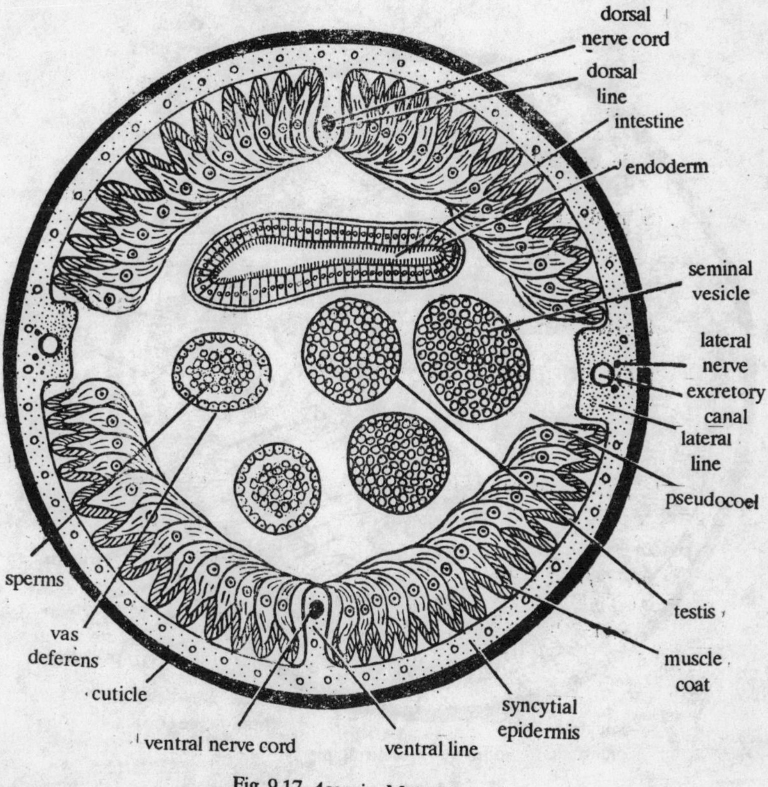

Fig. 9.17. *Ascaris*. Male T.S.

II. Female Reproductive Organs

The female reproductive organs include :

1. A pair of ovaries
2. Oviducts
3. Uteri
4. Vagina

1. *Ovaries.* The ovaries represent *didelphic* condition i.e. there are two ovaries. Each ovary is a long, thread-like, much twisted and blind tubule occupying posterior 1/3rd of the body. Its walls consists of a single layer of cuboidal epithelial cells lined

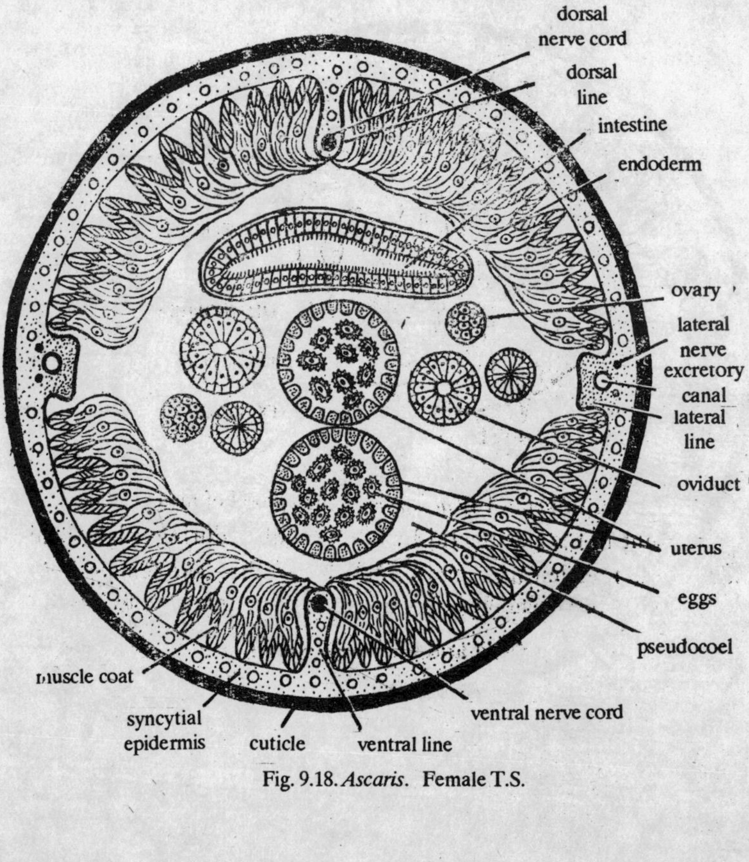

Fig. 9.18. *Ascaris*. Female T.S.

externally by basement membrane. There is no lumen, the central part is occupied by the cytoplasmic *rachis,* around which young eggs are clustered.

2. *Oviducts.* The oviducts are two thin ducts very similar to the ovaries. These open into the uteri.
3. *Uteri.* Each oviduct continues into a much wider, thicker and almost untwisted tube, the *uterus.* The wall of the uterus is composed of tufted cells, surrounded by a muscle layer with inner *circular* and outer *oblique* fibres.
4. *Vagina.* Two uteri join to form one vagina which opens to the outside by female genital pore or *vulva,* lying on mid-ventral side about 1/3 of the length of the body posterior to anterior end.

Life-cycle

Ascaris is *monogenetic* i.e., it requires only one host to complete its life-cycle.

1. *Copulation and Fertilization. Ascaris* copulate in the intestine of man. During the process sperms are ejected by male into vagina of the female. These move in the uterus where fertilization occurs. While moving further the eggs become covered by a chitinous shell and an outer irregular *albuminous coat.*

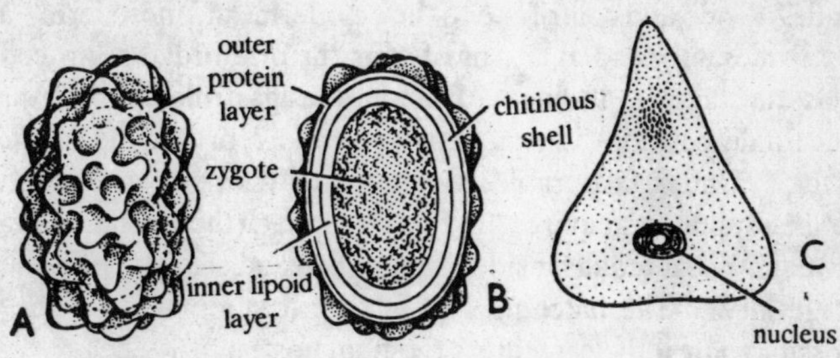

Fig. 9.19. *Ascaris.* A-egg, B-insection, C-a sperm.

The eggs are round or oval (60µ X 45µ), light brown in colour and float in saturated solution of common salt. The unfertilized eggs laid by the mother may mix up with the fertilized eggs but the unfertilized eggs are larger (90µ X 45µ) and sink in saturated salt solution.

2. *Egg-laying.* The uteri of a single female may contain as many as 2,70,000,00 eggs (*Cram,* 1927), with an average daily production of about 2,00,000 eggs. The fertilized eggs are released in the intestine of host through the vulva. From the intestine of host they are eleminated along with faecal matters.
3. *Development.* The development takes place at 85° F i.e. outside the body of man. It is stopped at 60° F and egg gradually degenerated at 100° F. They can remain alive for years in moist soil.

Cleavage. The cleavage is spiral and determinate. The first division is transverse producing an upper *dorsal cell* (*AB*) and a lower *ventral cell* (*P*). The dorsal cell (*AB*) divides vertically into an *anterior cell* (*A*) and a *posterior cell* (*B*). Simultaneously, the ventral cell divides horizontally into *upper cell* (*EMST*) and *lower cell* (*P*). Thus the four-celled embryo has *T*-shaped appearance. The P_2 cell shifts towards right and comes to lie in front of the upper cell (*EMST*). The *A* and *B* cells divide together into right and left cells and by further divisions form the ectoderm of the embryo. The *EMST* and P_2 cells divide into *E, EMST,* P_3 and *C* cells. The offsprings of cell E form the endoderm, while the descendants of *EMST* constitute the mesoderm and a part of ectoderm. The cell P_3 divides into P_4 and *D*. The cells *C* and *D* divide and contribute to the ectoderm and mesoderm. The cell P_1 divides into G_1 and G_2 and forms the primordial germ cells. Thus blastula stage is attained. It undergoes gastrulation by invagination and finally develops into an active juvenile, representing the *rhabditoid stage.* Under favourable conditions of temperature, moisture and oxygen supply, it requires 10-14 days to reach the 1st stage larva. After a week or so it moults inside the egg shell and changes into the second stage larva. The hatching and further development of the larva is possible only if the egg enters the final host.

Infection and Emergence of Larva. The eggs containing rhabditiform larva contaminate food and water. If someone takes this contaminated food or drinks contaminated water the eggs reach the healthy man. This type of *transference* is called contaminative transference or infection. The shell of the egg is weakened in oesophagus and is dissolved in the intestine of the new host. The larva is about 200-300u in length and 14u in breadth. It has fully formed alimentary canal, excretory and nervous system.

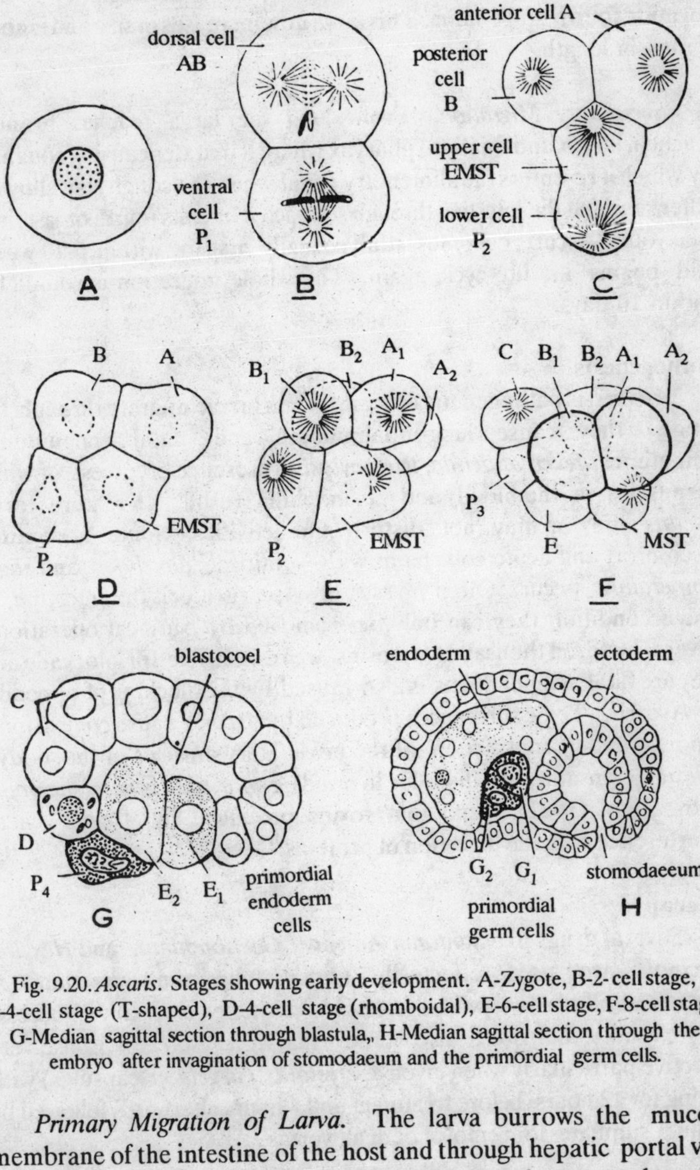

Fig. 9.20. *Ascaris*. Stages showing early development. A-Zygote, B-2- cell stage, C-4-cell stage (T-shaped), D-4-cell stage (rhomboidal), E-6-cell stage, F-8-cell stage, G-Median sagittal section through blastula,, H-Median sagittal section through the embryo after invagination of stomodaeum and the primordial germ cells.

Primary Migration of Larva. The larva burrows the mucous membrane of the intestine of the host and through hepatic portal vein reaches the liver. From here through portal vein and post-caval reaches the right auricle of the heart. From here through pulomonary arteries reaches lungs. In lungs it moults twice. The first moulting takes place after 5-6 days (third stage larva) and second moulting after 10 days forming fourth stage larva. This larva now increases in size and is about 2 mm. in length.

Secondary Migration. From lung the larva reaches bronchi, trachea, larynx and finally to pharynx where it irritates causing coughing by which it re-enters the alimentary canal when the cough is swallowed. After reaching the intestine through stomach it moults fourth or last time. The young Ascaris becomes adult sexually mature within 6-10 weeks and begins its life-cycle again. The whole migration is completed within 10 days.

Pathogenesis

Ascaris is pathogenetic to man. When larvae migrate through the lungs, they cause lung-inflammation and fatal pneumonia. Sometimes, *fever, anaemia, leucocylosis* (presence of excess of white corpuscles in the blood) and eosinophilia result. In mature form, *Ascaris* may or may not disturb life-activities. Some abdominal discomfort and acute colic pains with vomitting, *diarrhoea* and *slight temperature* occur. Often masses of Ascaris block the intestine, in which condition they can only be removed by surgical operations. Several hundred thousands of worms are responsible for blockade and they are fatal. Appendicitis is also caused by the blocking of appendix by Ascaris. Toxic substances produced by *Ascaris* cause *convulsions,* (an involuntary contraction of the involuntary muscles of the body), *delirium,* (insanity or disorder in mind), *coma* and *general nervous-ness. Sang* (1922) found that toxins produced by *Ascaris* cause interference with the digestion of proteins.

Therapy.

Several drugs like *Santonin Antepar, Chenopodium* and *Hexylre-sorcinol* are used effectively under physician's directions. A mixture of oil of *chenopdium tetrachlorothylene* is usually success-fully used. *Lamson* discovered that *hyxylresorcinol* is also very effective particularly when given as *crystoids* (in *gelatin* capsules) with fasting for 12 hours before treatment and 4 hours afterwards followed by sodium sulphate for removal of deal worms.

Hetrazan syrup is also very effective and nontoxic especially for children. *Piperazine hydrate,* Alcopar and Banocida are also recommended by several workers as antihelminthes.

Prevention.

Adequate sanitation plus education should prevent ascariasis. Lack of adequate fertilizer in many parts of the undeveloped world means that night soil must be used to fertilize crops. Due to the resistance of *Ascaris* eggs to standard disinfectants, the situation would appear to be hopeless. There is a realistic solution through; *Thitasut* (1961) reported that aqueous solution containing 120 ppm free iodine killed *Ascaris lumbricoides* eggs within 5 min or less and did not effect the flavour of leafy vegetables.

One of the problems in control is preventing children from ingesting eggs with contaminated soil. given the habits of children throughout the world, including dirt eating, playing in contaminated soil and then placing their fingers in their mouths, ascariasis will unfortunately be with us until adequate sanitation is accomplished.

10

ANCYLOSTOMA

No group of the nematodes causes more injury to man or greater economic loss through attacks on his domestic animals then the members of the order Strongyloidea, the great majority of which are parasites of mamamals. Many are blood suckers and cause severe injury to their hosts by loss of blood sucked by them or wasted from hemorshages, the result is anemia.

SYSTEMATIC POSITION

Phylum - **Aschelminthes**
Class - **Nematoda**
Order - **Strongyloidea**
Genus - ***Ancylostoma***

Distribution

Human hookworm is endemic in warm, wet climates, but each species has some limitation on its distribution.

Ancylostoma duodenale is the common hookwork of man in tropical, subtropical and temperate regions of Asia, Africa, Pacific islands and South America. The other important human hookwork is *Necator americanus,* more common in the United States. Some other species of *Ancylostoma,* namely, *A.caninum* and *A.braziliense,* both in dogs and cats, and *A. malayanum* in bear, are also common.

Habit

A. duodenale is the principal hookworm that infect humans. Following early morning, they are found in the small intestive lumen.

Hookworms sometimes mature in abnormal hosts, and hosts other than man can be infected with human hookworms.

Morphology

The name 'hookworm', is rather misleading because there is absence of hooks. The anterior hook-like bend probably is responsible for the common name, hookworm.

Shape and Size. The sexes are separate. It is a cylindrical worm and whitish or grey in colour. The females are longer than males measuring 10-13 mm. in length and 0.6 mm in width. The males measure 8-11 mm. in length and 0.45 mm. in width.

Buccal capsule. The anterior end of male and female is similar, somewhat narrower and curved dorsally. It bears a large cup-shaped buccal capsule for attachment to the intestinal wall. The buccal capsule has two *chitinous plates*, bearing teeth. The *teeth* make a tiny cut in the intestinal wall of the host and hook-worm sucks blood by the sucking action of pharynx.

Copulatory bursa. The posterior end of female is somewhat taper, bluntly rounded in the form of a cone and terminates into a minute spine.

The *posterior end* of male is broad and with an umbrella-like structure called *bursa.* The bursa is supported by fleshy bursal rays of definite arrangement.

Digestive System. The digestive system is a straight tube comprising mouth, buccal capsule, pharynx, intestine, rectum and anus.

The phanynx has triradiate lumen lined with in turned cuticule and a wall composed of strong radiating muscle fibres extends posteriorly from the buccal capsule about 1/6 the body length. The posterior region of pharynx is slightly dilated and embedded in its wall are three pharyngeal glands. One is median dorsal in postion and a pair located subventrally. These glands open into the pharyns near the nerve ring. An intestive begins immediately behind the triradiate pharynglal valve and proceeds as a straight tube to a short recturm. It is composed of a single layer of columnar cells. In the intestive the blood of host and other tissues are digested.

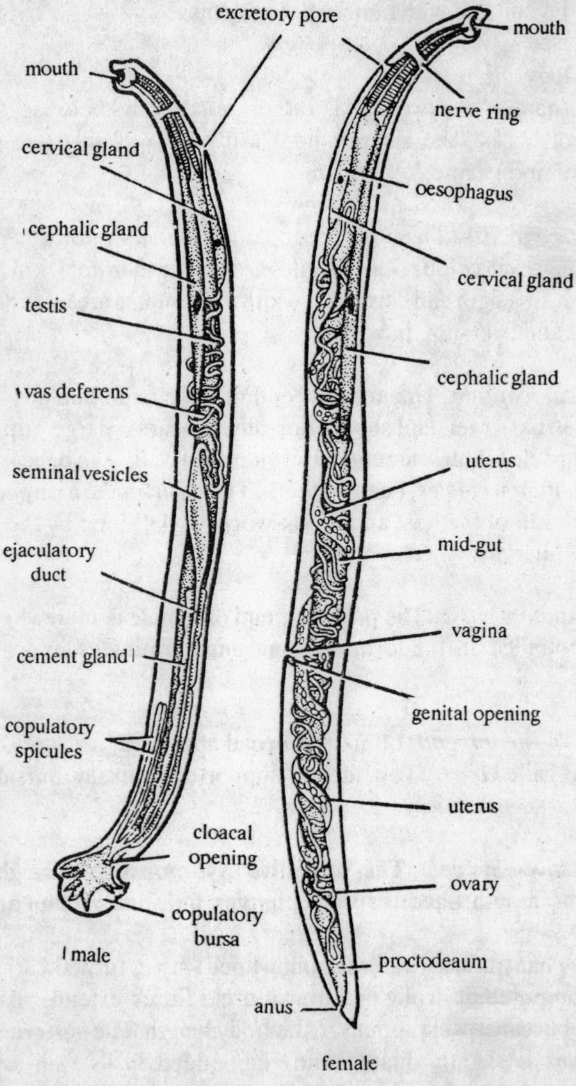

Fig. 10.1. *Ancytostoma duodenale.*

Musculature. The somatic muslces consists of a single layer of cells underlying a hypodermis and arranged into four longitubinal groups. These four groups are interrupted by four longitudianl quadratal chords located on the dorsoventral and leteral axes. Because muscle cells are limited to two in each sectioned quadrant, the worms are called *meromyarial.* The almsot complete lock of transverse muscle cells.

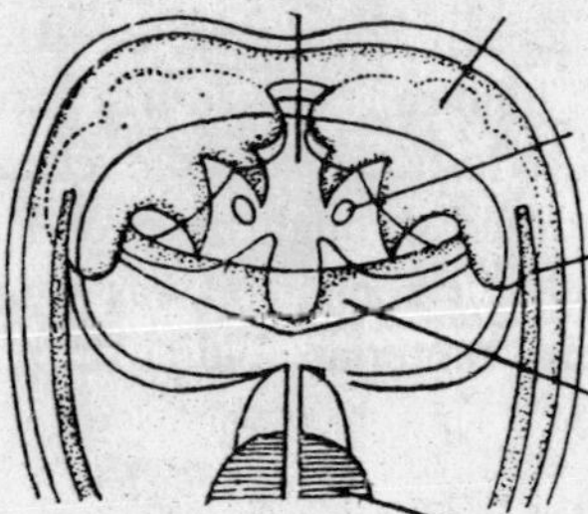

Fig. 10.2. *Ancytostoma*. Buccal capsule.

Excretory system. The exretory system is rather complicated. Flame cells are absent in nematodes. The excretory system, consists of two canals embedded in the lateral chords and an isthums in the pharyngeal region. Surrounding or adjacent to the insthums, are other excretory structures including an excretory sinus into which tributaries from the lateral canals open and from which ventrally is given off a short, cuticulate, terminal duct. The duct opens at a ventral exretory pore. From the excretory sinus two large, subventral glands of uncertain function arise posteriorly. The tributory canals emerge from subventral glands and penetrate the lateral chords where they join the lateral canals. These lateral canals extend from the region of buccal capsule to the posterior end of the body whose they end blindly.

Nervous system. Encircling the pharynx, in front of the exretory pore, is a nerve ring. From a large ganglion located on the ventral portion of the nerve ring, a large nerve runs in the ventral chord to the posterior end, and smaller nerves run in the lateral chords anteriorly and posteriorly. Four small nerves run forward from the ring to supply anterior end. These main nerve trunks are connected in a complicated fashion by nerve commissures.

Reproductive System

The male and female reproductive system are as follow :

Male reproductive System. The tubular testis occupies the middle third of the body. It begins at the junction of the middle and posterior body thirds and runs in a random fashion to the middle of the body, from which point forward it coils loosely about the intestive. It leads posteriorly into vas deferens followed by seminal vesicle. Seminal vesicle which connects posteriorly to an ejaculatory duct leading directly to the cloaca. The ejaculatory duct, through most of its course, lies between two large, multicellular cement glands whose secretion apparently serves as an adhesive moterial holding the pair of worms together at copulation.

The posterior end of male bears a wide bursa, two copulatory spicules and a gobernaculum. The bursa is a long, wide, bilateraly. Symmetrical structure, with supporting, paired, fleshy rays having a specific distribution. The two spicules are long and slender. Each measures 2000 um. They ae proturded and retracted by special muscles and help in the transport of sperms to the vulva of female. The gubernacculum is a hard peice of on the dorsal wall of cloaca. It is said to be used to guide the spicules.

Female reproductive System. Two much convoluted orarian tubules, one anterior and other posterior to the level of gonopore are present. From the ends of the ovares nearest the vulva short, narrow oviducts carry the ova into wider seminal receptacles where fertilization occur. From each seminal receptacle arises a muscular uterus that terminate into vagins. Two vagine join and then open to the surface at the vulva.

Life-cycle.

Copulation and fertilization. Fertilization is internal and during copulation the bursa of male is applied on the vulval of female. The female Ancylostoma lays about 20,000 eggs per day. The egg is about 60 x 30μ. These are passed along the host faeces at four-celled stage and do not develop further until exposed to air. Under favourable conditions the embryo hatches out of the egg in less 24 hours in the faeces contained in the mud.

Larval development. Under favourable conditions of moisture, oxygen supply and temperature (68-85°F) the embryo develops and hatched as larva.

The hatched larva is the first larval stage or the fist *'rhabditiform larva'*. It has mouth, pharynx and a simple intestine. It feeds on bacteria and organic debris in the faeces and grows rapidly. At the end of 2nd day it undergoes first moult and changes into second stage larva (or the *strongyliform larva*). At about 5th day it undergoes second moult to become *filariform-juvenile.* This is the infective non-feeding stage. The skin of the second stage larva is retained as a protective sheat. The larvae live in the upper half inch of soil.

Infection. The third stage or the filariform juvenile represents the infective stage. It may remain alive in the moist soil for nearly four months without feeding. When a man walks bare-footed on the infested soil, the larva penetrates the skin. Infection can also occur by ingestion of infective juvenile in food or drink.

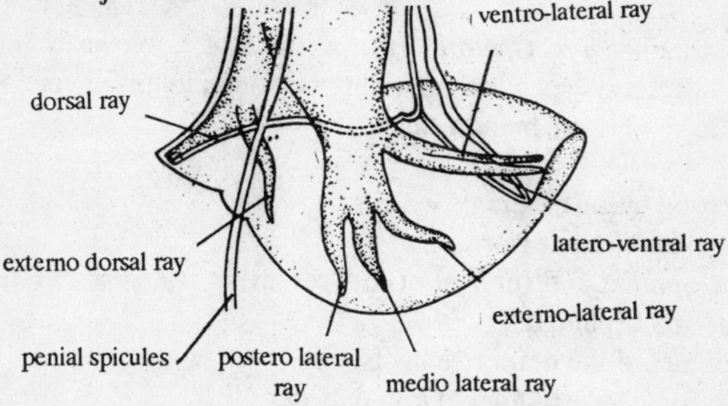

Fig. 10.3. *Ancytostoma*. Copulatory bursa.

Larval Migration. Within 24 hours. of infection, the larval reach the blood-vessels. From here they are carried first to the right ventricle of the hearst and then to the lungs. From lungs the larval assend the trachea and to the pahrynx. In pharynx they are swallowed through oesphegus and finally reach to the small intestive. In the ileum they become attached to the mucous lining and feed on blood of the host. In 5-6 weeks, they moult twice to become adult and begin to reproduce. The normal life-span of hookworm is 5-7 years.

Pathogenecity. Hookworm is the most pathogenic parasite of man.

1. *In skin* - The pentrating larva causes 'ground itching' or 'water sore' characterised by litching the inflammation of skin.
2. *In lungs* - The migratory larvae when in large number cause haemorrhage and branchial pneumonitis.
3. *In intestine* - Adults in the intestine produce stomach pain, abdominal discomfort, nausea, food fermentration, diarrhoea. But the principal effects are due to anemia, resulting from constant sucking of blood. It leads to palpitation of heart, eosinophilia and sucking of blood. It leads to palpitation of heart, eosinophilia and even collapse. Hookworm infection in children causes retarded mental and physical growth and sometimes to dizziness, insomnia, optical illusions, general nervousness and fidgety movements.

Treatment. A number of highly effective antihelmintics are available for use in humans, including piperazine, bengimidazoles, thymol, carbontetrachloride and Hexlyresorcinal.

Epidemiology *and Control.* Transmission of hookworms in the buman population is dependent on a combination of following biological and climatic factors and human culture.

1. Egg production of female worms.
2. Longevity of adult worms.
3. Development and survival of the free-living states as related to weather and climate.
4. Immunity of the definitive hosts.
5. Nutritional level of human population.
6. Technological and economic level of the area.

A total compaign against hookworm disease involves :-

1. Mass chaemotheraphy.
2. Samitary disposal of human waste.
3. Wearing shoes.
4. Administration of ions salts.
5. Ensuring adequate dietary protein.

11

WUCHERERIA (= FILARIA) BANCROFTI

The parasite gained importance after *Manson's* findings in 1878 that a mosquito acts as the transmitted agent for this parasite. *Manson* (1887) studied the life-cycle of the parasite and suggested reasons for the marked diurnal periodicity exhibited by filaria in the peripheral human circulation. *Vogel* (1928) and *Fain* (1951) gave accounts of the adult structure. The importance of this parasite to mankind is underlined by the fact four international meets were held in late 1960's to discuss the epidemiology, pathology and control of this worm.

SYSTEMATIC POSITION

Phylum - **Aschelminthes**
Class - **Nematoda**
Sub-class - **Phasmida**
Order - **Filaroidea**
Genus - ***Filaria***
Species - ***bancrofti***

Distribution

Filaria, the most important human filarial parasite, is widely distributed in a tropical and subtropical circumglobular belt. But the parasite is not evenly distributed or uniformly prevalent throughout any countary. It occurs almost entirely in costal areas and islands where there is a fairly long hot season with high humidity. In Africa found on the Mediterranean and east and west costal areas but not in the interior of Central Africa. In Asia it is prevalent on the coasts of Arabia, India, Malaya, Formosa, and north to China and the southern part of Korea and

Japan. It is prevalent in practically all the East Indian and South Pacific Island.

Habit

It is a *digentic parasite* i.e. its life-cycle is completed in two hosts. The *definitive host* in man where it lives in the lymphatic system. The *intermediate host* is mosquito (*Culex, Aedes* and *Anophiales*).

Morphology

The adult worms are found in the lymphatic vessels and lymph nodes of man only.

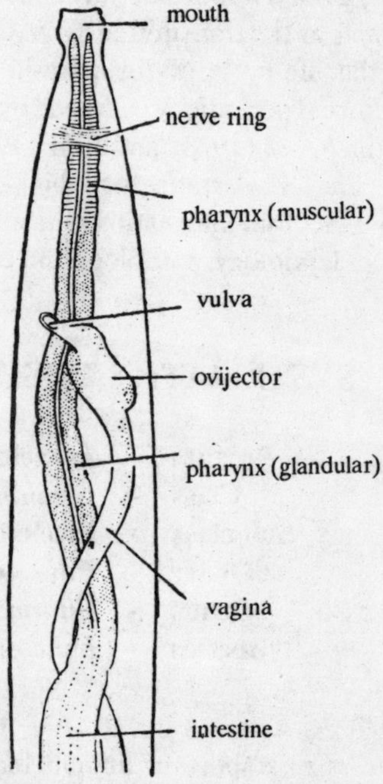

Fig. 11.1. *Wuchereria bancrofti.* Anterior part

Adult worms are hair like, transparent nematodes often creamy white in colour. They are filiform and both ends are tapering. The posterior end terminates in a slightly rounded swelling. The male is about 40 mm, in length while thickness is about 0.1 mm. The posterior end of the male is curved ventrally and has two spicules of unequal length. Female measures 65-100 mm, in length and its thickness is about 0.2 to 0.3 mm. The tail end of the female is narrow unbent and abruptly pointed. Although females, liberate active embryos yet they are really *ovo-viviparous* i.e., lay eggs with well developed embroys. Males and females usually remain coiled and can only be separated with great difficulty. Females are more in number than males. The life-span of the adult worm is about 4-5 years.

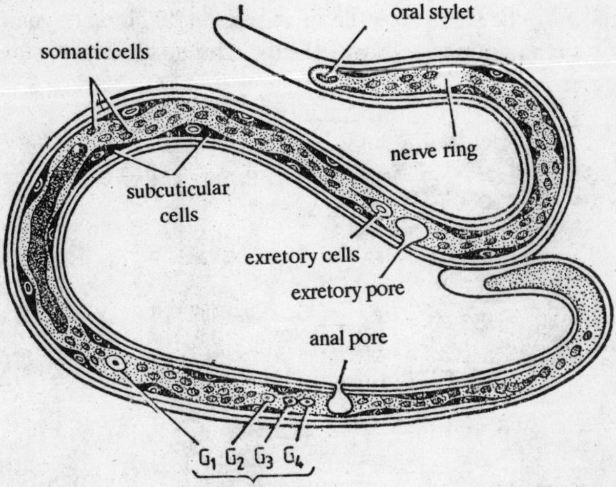

Fig. 11.2. *Wuchereria*. Microfilaria.

Life-Cycle

Copulation takes place when individuals of both the sexes are present in the same lymph node. Female, releases numerous (about 1,000) first stage larval or jureinule called *microfilariae*. These larvae are surrounded by delicate membranes or sheaths and measure 225-300 mu long and 10mu in diameter. When living they are colourless and transparent and may or may not be surrounded by sheaths. In order to identify them they are usually stained. The body will the be found to contain a column of nuclei, broken in definite places. The peripheral ores are a nerve ring anteriorly, an excretory pore or 'V' spot, an excretory call somewhat further back, a few genital calls posteriorly and an and pore. The tail is devoid of nuclei.

The microfilairal appear in the peripheral circulation chiefly between 10 P.M. and 4 A.M. There has been much spaculation and experiment to determine the reason for this periodicity. According to one view, the larval are concentrated in internal organs such as lungs (Smyth, 1976) during day time, but they circulate in the blood at night for the transmission to the mosquitoes. Another view holds that embryos are born at a certain time each day and are then destroyed in the host within next 24 hours. According to *Manson-Bahr* and *Muggleton,* the nocturnal periodicity of mcrofilarial in peripheral blood of man is an adaptation to maintain a tryst with the night biting species of mosquito.

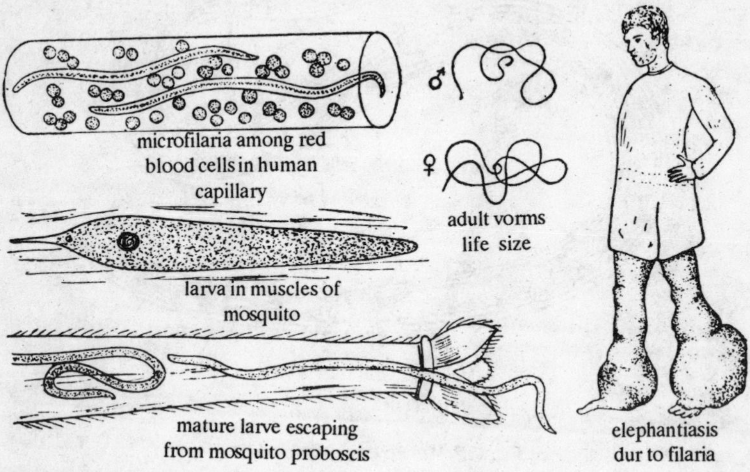

Fig. 11.3. *Wüchereria*. Stages in the life-cycle.

The further development of the microfilarial depend on their being sucked with blood by certain species of mosquitoes. In order to infect mosquitoes there must be about 15 or more microfilarial per drop of blood, a high concentration of 100 or more per drop of blood is fatal to mosquitoes.

Shortly after being digested by a mosquito the microfilarial penetrate through the stomach wall and migrate to the breast muscles. Here the embryos lie length wise between the muscle bibres. The size of body is reduced to half it original size length, but grows several times as thick. It now changes into a sausage-shaped creature.

The sausage shaped larva measures 1.5-2.0 mm. in length and 20-30 mµ in diameter. As these changes take place the larva mouths twice and become infective. The formation of the infective stage takes about 10-14 days and occurs best at 28°C and a relative humadity 90. Digestive system, body cavity and genital organs also become well developed. This is called third stage larva. This infective stage reaches the into the head of mosquito and finally into the proboscis on about 14th day. There are several larvae, coiled up in mosquito ready for infection in man. The complete development of microfilaria in the mosquito is from 10-20 days.

Infection and adult stage in man. When the infected mosquito bites a man, the third stage larvae are deposited, usually in pairs, on skin near the site of puncture. Later on the larvae enter through the puncutre or penentrate through the skin. From here they reach the lymphatic vessels and settle down at same spot, i.e. inguinal, scrotal or abdominal lymphatics. The larva grows into adult. After a period of 5-18 months, larval become sexually mature and start new generation of microfilarial.

Pathogenic effect

The pathogenic effect of *Filaria* produced by the adult *Filaria* either living or dead. When microfilariae are circulating in the blood, no pathogenic effects are produced. The adult worm causes an inflamatory reaction of the lymphatic system or *lymphangitis.* The metabolities of the growing larvae cause allergic manifestations such as urticaria, fugitive swellings i.e., raised, painful, tender red areas of the skin at extremities and *lymphoedema.*

The cause of lymphangitis are mechanical irrigation, liberation of metabolites, absorption of toxic products, and bacterial infection. Later when lymphetics passage are severly blocked by the adult and microfilariae *elephentiasis* results.

Treatment

Drugs for *filaricidal* reactions can be grouped into 3 main groups :

(1) on adult worms-Mel w, an arsenical preparation.
(2) on microfilariae-Diethycarbamazine (Hetrazan).
(3) on infective larvae and immature adult worm-para-melaminyl-phenyl stibonate (MSb).

Prophylaxis

It includes :

1. Killing the adult intermediate host i.e. mosquitoes.
2. Use of selected insecticides to kill the adult and larval stages of mosquito.
3. Use of wire netting at door and windows of human dwellings to prevent the entry of mosquitoes.
4. Administration of diethylcarbamagine (DEC) mixed in common salt has shown encouraging results.

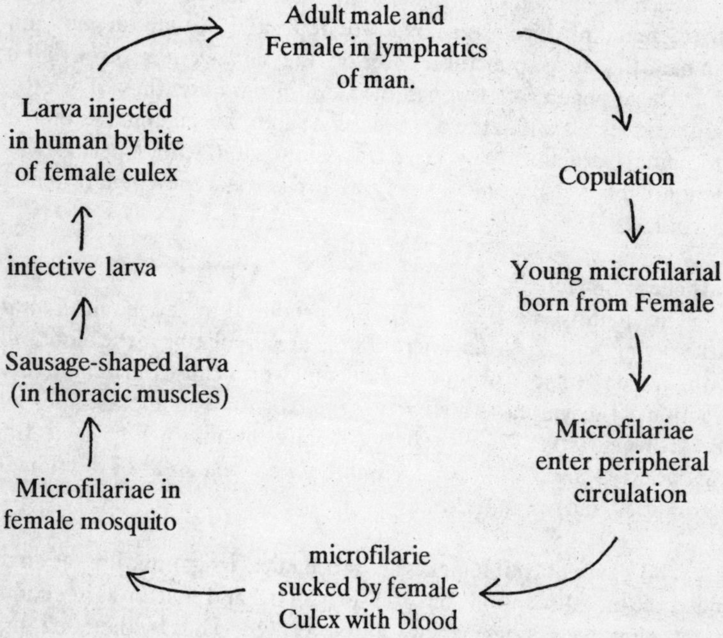

Life-cycle of *Filaria bancrofti*.

12

PARASITISM AND PARASITIC ADAPTATIONS

Parasitism is such an association between two animals of the same or different species in which one lives on or inside (parasite) the body of the other (host) receiving nourishment and shelter without any compensation for the host. Therefore, in this association the parasite is benefited and the host is harmed.

Origin and evolution of parasitism

Parasitism is a secondary mode of life and his arisen from pre-living condition. Parasites living inside the bodies of their hosts (*endoparasites*) are the most successful ones. Between the endoparasitic life and free-living life, intermediate stages are not uncommon, proving that parasitism did not evolve all of a sudden but followed a series of steps in succession.

Living together or symbiosis started when smaller animals occasionally taking shelter temporarily on the body of some larger form. When this temporary shelter was repeatedly used, the association of two organisms become more intimate and the small organism got not only the shelter but also free transport and nourishment in the form of bits from the host, without interfering with the activities of the host or injuring it. Such a simple symbiosis is known as *commensalism* and the benefactor is called a *commensal.*

Still later, the commensal started living within the body of the host without doing any harm to the host. This a commensal may be firstly an *ectocommensal* and later an endocommensal. Symbiosis sometimes proves beneficial to the host. Thus both the organisms benefitting each other mutually the host becomes so dependent upon the commensal that it cannot get along without it. This kind of association is called *mutualism.*

Commeusalism usually evolves not in the direction of mutualism but towards a deeper association, the *parasitism*. A commensal that at first takes shelter, then bits of food and finally starts to feed on body tissues of the host so that the latter has to suffer with some harm.

Different types of parasitism

The parasitism can be classified into following catagories :

1. *Accidental*. When a free living animal accidently reaches inside the body of the host then it leads a parasitic life for some time.
2. *Facultative*. When the animal can live both as parasitic or free-living, they are known as facultative parasites.
3. *Obligatory*. These animals can not lead a free life, therefore, they need a suitable host for them. The obligatory parasitism is of different types :

(a) *Ectoparasitism*. The animal lives on the external surface of the body of the host.

(b) *Endoparasitism*. The animal lives inside the the body of the host in various organs and systems.

Effect of parasitism upon the parasite.

Changes from a free-living condition to a parasitic life involves changes in the body of the animal. Due to parasitic mode of life certain new structures are developed due to new environmental requirements similarly certain characters, which are of no use for new environment, and lost. All such changes are included with effect on parasitism on the parasite. The helminths are modified morphologically as well as physiologically to live in their particular environment.

I. Morphological Adaptations.
II. Physiological Adaptations.

I. Morphological Adaptations.

These are of two types (i) Structure which are partially or wholly lost. (ii) Structures which have developed a fresh.

Degenerations.

1. *Organs of locomotion*. Locomotion is actually an effort for procuring food. But parasites habitually inhabit such places in the host's body, where sufficient food is available without effort. Thus, the organs of locomotion, such as the cilia of free-living turbellarians, are absent in the parasitic forms. It is interesting to note that the locomotory organs are duly present in free-living larvae of parasitic forms; the miracidium possesses cilia and the cercaria possesses a tail for locomotion.
2. *Organs of nutrition*. Food of the parasite comprises the readily available digested or semidigested food of the host. Therefore, there is a reduction in trophic organs. In trematodes, there is an incomplete gut and in most of them, suctorial pharynx is present. In cestodes, the mouth and alimentry canal are lacking because digested food is readily available in the gut of the host.
3. *Nervous System*. Living safely in a more or less stable environments, parasites do not need a complex nervous system.
4. *Sense Organs*. Progressive parasitism is accomplished by the loss of sense organs. The sense organs, necessary for quick and efficient response to the stimuli, are associated with the free active life. In the host the environment is more or less uniform and so the sense organs are not essential. Hence they are reduced. The endoparasites are generally provided with tangoreceptors.

Neoformations.

1. *Protective covering*. There are no epidermal cells in the adult, instead of which the body is covered by a several-layered thick cuticle which protects the parasite from the juices of the host.

The outer surface of the cuticle is formed of a fibrous protein called *keratin*. The cuticle may provided with scales.

2. *Organs of attachment*. For a firm grip on or in the host's body, some special organs of adhesion are necessary. The flatworms, for this purpose, are variously armed with suckers, hooks and spines. The suckers themselves may be with or without hooks and spines
3. *Organs for penetration*. The parasites must penetrate the host's body for entering it. So the parasites develop certain structures for

penetration. Miracidium larva has a conical process at the anterior end called *apical papilla.* These are a pair of penetration glands inside the body near the anterior end.

4. *Suctorical pharynx.* All the endoparasites except cestodes have a suctorial pharynx by means of which liquid food is sucked into the intestine. Intestine is always branched.
5. *Reproductive system.* It is highly developed and the production of eggs is prolific. The parastic flatworms are hermaphrodite. In cestodes the reproductive organs are much more elaborate and are repeated in each proglottid. Each mature proglottid possesses one (*Taenia sollium*) or two sets (*Diphylidium*) of male and female reproductive organs. In each gravid proglottid all other organs of the system degenerate to make room for the uterus which is greatly enlarged and branched to accomodate a large number of eggs.

II. Physiological adaptations.

1. *Protective mechanism.* In the alimentry canal of the host the parasite protect itself from the action of digestive juice of the host.

   The tapeworms accomplish this (a) by stimulating the walls of the gut to secrete mucus, which then forms a protective clothing around the parasite, (b) by secreting antienzymes to neutralize the digestive enzymes of the host, and (c) by probably continually renewing their protective body covering *i.e.*, tegument. Lime cells in the body wall of tapeworms nuteralize the acids produced by the host. Blood parasites are known to withstand the effect of antibodies and phagocytes of blood by some mechanism.

2. *Osmoregulation.* The Osmotic pressure of body-fluid of endoparasites, especially in case of trematodes, is almost the same as that of the host so that there is no necessity of osmoregulation. But in the intestinal tapeworms, the osmotic pressure is slightly higher than that of surrounding medium. This permits absorption of nourishment from their general body surface from the surrounding medium.
3. *Anaerobiosis.* The pH tolerance of the parasite is high from 4-11. The intestinal parasites live in an environment completely devoid of oxygen. The respiration is of the anaerobic type consisting of extracting oxygen from the food stuffs. In the absence of oxygen energy is obtained by the fermentation of glycogen in which glucose in broken into lactie acid.

4. *Periodic appearence.* There are some parasite which appears at definite period. The larval of *Wuchereria bancrofti* circulate into the peripheral blood circulation in the night between 10 P.M. to 4 A.M. for further development they require *Culex* mosquitoes. The *Culex* bites a man at night *i.e.* nocturnal in habit. Their migration is correlated with the nocturnal habit of *Culex*.
5. *Neoteny.* It is a special phenomenon in which the larval are capable of reproduction and they produce young ones. *Ligula* (Cestoda) exhibit neoteny and this can be considered an adaptation for parasitic mode of life.
6. *High fertility.* Parasites develop enormous fertility, producing millions of fertile eggs. This is correlated with the passive transference of the infective stages of the parasites from one host into another. While passing through the complex life-cycle, these potential offsprings face several hazards as a result of which a very small percentage of the total eggs produced reaches adulthood. For example in *Fasciola*, if the eggs fall in water, if the first larva (miracidium) finds a suitable snail within a limited period, if the cercaria encyst, on vegetation within the reach of final host and if the metacercaria happen to be ingested by the proper final host, only then the life-cycle is completed. Not only this, the larval stages themselves multiply say for example for sporocyst produces a number of rediae. A single redia produces a number of cercaria larvae. Similarly a single hexacanth may multiply by generating buds which produce daughter and grand daughter scolices (hydratids of *Echinococcus*).
7. *Transferance of eggs or infective stages.* This is another major problem for which the parasites have to develop adaptations. Transfarence from one host to another is either :

(i) *Active transfer.* This is a less common method. The infective stages are actively transferred to final host. The larval stages of most parasite bore directly through the skin of their host such as the larvae of hookworms and schistosomes.

(ii) *Passive transfer.* This is the most common method and usually results from the careless habits of the host and is achieved in several ways such as contamination, contagion and inoculation. Oral infection occur when the host takes uncooked food or contaminated water. The intermediate host, such as mosquitoes, inoculate parasites when they suck the blood of hosts.

Effect of parasitism on the host.

Although a given host may he immiene to a parasite by virtue of being constitutionally unstable as a habitat, it is important to recognize that the establishment of a parasite in a host also involve the capacity of the parasite to resist reactions of the host that might be deleterious. These host reactions would include inflammatory tissue responses and immume responses of a humoral character. Both types of host reaction occur in most parasitisms involving characters. The immune system in higher vertebrates is composed of several tissue components. These include fixed cells of spleen, liver, lymph nodes and thymus. There is also a system of mobile phagocytic cells, originating mainly in the bone marrow including the heterophil and eosinophil leucocytes. In addition, there are a number of potential macrophage cells in the body, such as undifferentiated mesenchyme cells. Some of these cells have the ability to elaborate antibodies, which may be released from cells into the body fluid. These globulins may combine with parasitic organisms or their products and produce an inhibitory effect. A peculiar type of immunity develops in man due to the presence of tapeworms; after the first infection no further infection is possible as long as the primary infection lasts, this is termed *premunition*.

The possible role of interferons in the immunity of hosts to animals parasites should be also mentioned. These substances are produced in response to the presence of a foreign agent. Unlike antibodies, they seem to render host cells until as habitations for intracellular parasites.

There are other types of host response that cannot be clearly recognized as immune in characters. For example, the changes in the fatty acids found in the alimentary canal of animals infected with the tapeworm, *Hymenolepis*, or the marked modification of the amino acid composition of bile in sheeps harboring *Fasciola hepatica* are obviously host responses to the presence of parasites concerned. It would be of great interest to determine whether such response are beneficial to the parasites, to the host or to neither.

INDEX